Homebrew Cookbook

By
Eamon Skelton, EI9GQ

Edited By
Elaine Richards, G4LFM

Radio Society of Great Britain

Published by the Radio Society of Great Britain, 3 Abbey Court,
Priory Business Park, Bedford MK44 3WH, United Kingdom

First published 2010

ISBN: 9781 9050 8657 3

Publisher's note
The opinions expressed in this book are those of the author and not necessarily those of the RSGB. While the information presented is believed to be correct, the author, the publisher and their agents cannot accept responsibility for consequences arising from any inaccuracies or omissions.

Editor: Elaine Richards, G4LFM
Cover design: Kim Meyern
Typography and design: Mark Pressland
Production: Mark Allgar, M1MPA

Printed in Great Britain by Page Bros of Norwich

Contents

Foreword

HOMEBREW radio is amateur radio slang for home made radio. Any radio equipment that you build yourself can be regarded as homebrew. A homebrew project might be as simple as a wire aerial or as complicated as a state-of-the-art, all band, all mode transceiver. Your homebrew project can be built from your own design or from a published design. You might choose to purchase a complete kit containing all components and detailed instructions or you may use components from the bottom of the junk-box. This is not a universally accepted definition of homebrew radio. A few amateurs will argue that to be 'truly homebrew' a project must be built from scratch using your own design. I have even heard one amateur say that a project can not really be considered homebrew unless the constructor uses only home made components! As I don't know how to go about making a point-contact transistor from a lump of coal, I will continue to use mass produced components for now!

Just about anybody can build radio equipment. All you need is enough interest and enthusiasm. Anybody with enough knowledge to pass the Intermediate exam certainly can be a homebrewer. You don't have to be a professional electronics engineer or some kind of electronics guru. I have seen many radio enthusiasts who started out with little or no knowledge of electronics and went on to build complete homebrew stations.

The most basic toolkit will contain a soldering iron, a multi-meter and a few hand tools. You will need screwdrivers, long nose pliers and a sharp scissors or side cutter. You will also need some solder and a desoldering tool or some desoldering braid. In addition to the tools, you will need a few standard components. Many retailers sell resistors, capacitors, diodes, transistors and other components in 'lucky bag' packs. Recovering components from scrapped equipment is a less expensive way of filling your component box. TV, video, hi-fi and other electronic equipment is packed with useful components. I have built complete receivers and transmitters using only components that have been pulled from dead equipment. VHF two-way radio equipment and dead CB rigs are a particularly good source of radio components. If you are starting from scratch, you can probably get all you need for much less than £50. If you already have most of the tools listed above, you might expect change from a fiver.

If you don't have a multi-meter, you will have to decide whether you want to buy an analogue meter or a digital one. Analogue meters are good at measuring changing values. I always reach for my analogue meter when adjusting a circuit that gives a peak or dip in voltage or current when adjusted. A DMM (digital multi-meter) is more accurate and more precise than a typical analogue meter. The DMM usually has a much higher input resistance than its analogue counterpart. A meter with a high input resistance will have much less effect on the circuit you are testing. When choosing an analogue meter, it is worth spending a little more to get an instrument with a sensitivity of 20kΩ per volt or more. Choosing a digital meter is a much easier job. I have found that even the cheapest DMMs will do an excellent job. My favourite DMM cost just over £3 from a well known German supermarket.

You don't really need to have more test equipment at this stage. While it would be nice to have a shack full of expensive test equipment, our simple toolkit is all we need for now. More sophisticated test equipment can be bought or built later on. Surplus test equipment can often be bought at radio rallies or on internet auction sites.

The less experienced constructor should start with a relatively simple project. It can be very discouraging for a newcomer when a circuit just refuses to work. With a more complicated project, there is a much greater chance of things going wrong. When you have successfully completed a few simple projects, you will have developed the skills needed to tackle a bigger project. You will also develop the perseverance to keep hacking away when you run into problems. All homebrewers will fall foul of Murphy's Law **[1]** from time to time.

Eamon Skelton, EI9GQ

References

[1] Murphy's law: 'Anything that can go wrong will go wrong'.

1

Construction Methods

INTRODUCTION: I am a big fan of dead-bug construction and other similar point-to-point wiring methods. There are quite a few variations on this method. Some constructors use a hacksaw blade to cut the copper layer of PCB into a grid of isolated square or rectangular pads. The component leads are soldered to the pads and short lengths of wire are used to make connections between pads. **Photo 1.1** shows a simple circuit that was built using this method.

This example uses single-sided PCB. The components can be mounted on the same side of the board as the copper pads. If you drill holes for the component leads, the components can also be mounted on the opposite side of the board. If double-sided PCB is used, one side can be used as a continuous ground plane.

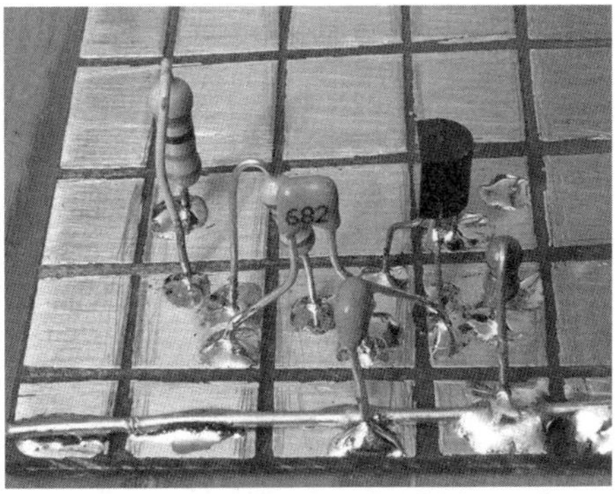

PHOTO 1.1: example of grid-based construction.

Another style of construction uses copper pads that are cut or punched from copper-clad PCB and glued to a copper ground plane. This method is known as 'Manhattan Construction'. Excellent Manhattan Construction tutorials can be found on K7QO's website **[2]**.

The methods described can produce attractive and functional circuits. I have found that the extra time and effort to cut out grids and pads can rarely be justified. Simple dead-bug construction is easier, quicker and just as reliable. To enhance the mechanical stability of a dead-bug circuit, some of the component leads can be soldered to PTFE or ceramic stand-off terminals.

PHOTO 1.2: VHF balanced mixer that uses press-in PTFE stand-off terminals.

Photo 1.2 shows a VHF balanced mixer that uses press-in PTFE stand-off terminals. Good old fashioned tag-strip terminals can also be used for this purpose. Very high value resistors can also be used as stand-off insulators. A 10MΩ resistor will usually have little or no effect on the operation of your circuit. One end of the resistor is soldered to the copper ground plane.

PHOTO 1.3: A 1296MHz band-pass-filter made from cut and glued line.

UHF and microwave circuitry often uses transmission lines as the reactive elements of filters and matching networks. The transmission line sections can be etched on double-sided PCB or cut from single sided PCB and glued to a copper ground plane. Air-dielectric lines can be cut from sheet metal and mounted a few mm above the ground plane. The simplest and most common arrangement is called 'microstripline'. This type of circuitry can easily be integrated with the more conventional dead-bug techniques we are already familiar with. 50Ω transmission lines are easily made from standard fibreglass PCB. Assuming that the board is 1.6mm thick and has a relative permittivity of 5.0, a 50Ω line will be 2.77mm wide **[3]**. This line can be etched on 1.6mm G10 or similar double-sided PCB or cut from single-sided PCB and glued to a copper ground plane. The cut-and-glue approach is more forgiving when you make a mistake. **Photo 1.3** shows a 1296MHz (23cm) band-pass-filter made from cut and glued line. Each of the three lines is shorter than a quarter wavelength and is tuned to resonance by a small trimmer capacitor.

Making Your Own PCBs

The printed circuit board (PCB), also known as printed wiring board (PWB), is quite literally the backbone of modern electronics. The vast majority of electronic circuits are built on printed circuit boards. The most common type of PCB has one or more layers of copper foil bonded to an insulating substrate. The substrate is usually made from a composite material like fibreglass/epoxy or SRBP (synthetic resin bonded paper). The copper tracks and pads are formed by chemical etching of the copper foil.

Single sided and double sided copper clad board is readily available from most component suppliers. Fibreglass board is a bit more expensive than SRBP, but it has a few advantages.

Fibreglass board is much more durable than SRBP. It also has more predictable electrical properties, is more flame retardant and is easier to cut with a shear or a saw. SRBP board is less expensive than fibreglass, and it is also much easier to drill. SRBP can be difficult to cut because it cracks very easily.

Most PCBs are made by drawing the track layout on copper clad board using an etch resistant coating and immersing the board in an etchant solution. Bare copper is removed by the etchant, leaving only the copper protected by the etch resist. After etching, the board is rinsed with clean water, the etch resist is removed and the board is thoroughly cleaned using a mild abrasive. Finally, holes are drilled for the component leads. The components can be mounted on the opposite side of the board to the copper tracks or surface-mounted on the same side as the copper tracks.

SAFETY. Please take great care when working with fibreglass PCB. Cutting fibreglass with a hacksaw produces quite a lot of fine dust. At the very least, you should wear a dust mask and work in a well ventilated area. Wear protective goggles when you are drilling. Don't leave PCB etchant or developer trays within reach of children or animals. Don't rub or splash glass fibre or chemicals into your eyes. Wear rubber gloves when working with PCB chemicals.

ARTWORK. There are many different ways of printing the track layout onto the copper PCB. The most obvious way is to draw the tracks by hand with an etch-resist pen. Special pens are available for this purpose. I have found that some brands of laundry marker make excellent etch-resist pens. Staedtler fine laundry markers give particularly good results. Tesco own-brand CD/DVD pens also make good etch-resist pens but I haven't had much success with ordinary waterproof/permanent felt-tip pens.

I etched a small scrap of PCB to test the effectiveness of a Staedtler fine laundry marker, a Dalo etch-resist pen and a Tesco CD/DVD pen. **Photo 1.4** shows the etched board. All three pens gave acceptable results. A close inspection of the board with a magnifying glass shows that the laundry marker wins by a small margin. It has slightly better etch resistance than the Tesco CD pen and can draw thinner lines than the Dalo. The Tesco pens are good value at four for less than a pound.

A scrap of Veroboard can be used as a template for drilling holes in hand-drawn PCBs. I have found that the best approach is to drill the holes for IC pads first. The holes can then be used as a guide for drawing the IC pads and tracks.

CAD/CAM. Back in the bad old days, PCB artwork was laid out by hand using rub-on tape and pads. These days, most PCB layouts are created on computers running CAD (computer aided design) software. PCB design software ranges from simple PCB CAD programs to complete EDA (electronic design automation) packages. If you don't need advanced features like schematic capture, circuit simulation and auto-routing, you could use a standard drawing program to create PCB artwork. I use a free and open-source PCB CAD program with the rather unimaginative name of *PCB* [4]. *PCB* runs on Linux and other Unix-based systems. There are several PCB CAD programs available for Windows and MacOS, some of which are available as freeware. Simple PCB work is not very demanding in terms of memory and CPU cycles, so a fairly humble PC is adequate for this task.

PHOTO 1.4: A comparison of the effectiveness of three pens for use as an etch resist marker
*** Staedtler fine laundry marker**
*** Dalo etch-resist pen**
*** Tesco CD/DVD pen**

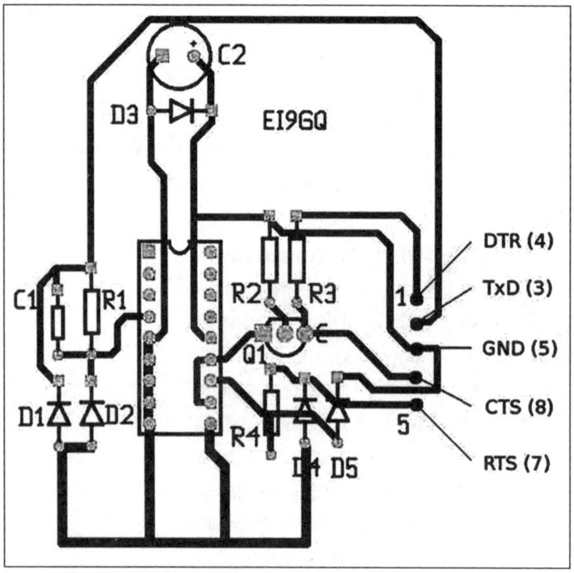

FIGURE 1.1: A typical layout for a simple single-sided PCB.

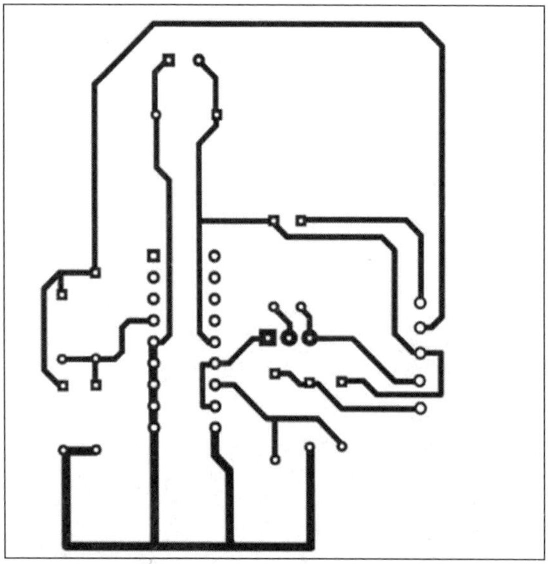

FIGURE 1.2: CAD-produced artwork used to create the PCB. Note that this is not to scale. Correctly scaled artwork can be downloaded from: http://homepage.eircom.net/~ei9gq/art.html

Figure 1.1 shows a typical layout for a simple single-sided PCB. The different colours show the various layers of the PCB. Even this simple example has a silk screen layer for printing the component outlines and names on the component side of the board, a solder resist layer that we won't be using and the copper track layer at the bottom of the PCB. Before the advent of surface-mount technology, single-sided PCBs usually had the components mounted on the top of the board and the copper tracks on the opposite side. Life was simple, the top was at the top, the bottom was at the bottom and everybody knew which was which. Boards using surface-mount technology (SMT) have the components mounted on the same side of the board as the copper tracks. Many PCBs have at least two copper layers.

Sometimes surface-mounted components will be mounted on both sides of the PCB or a mix of SMT and leaded components will be used. Complex boards are often made from a multilayer sandwich of copper and fibreglass. Such boards can have a ground plane layer, one or more power layers and several signal layers. Add a few more layers for solder mask and silk screen printing on both sides of the board and you will understand why some PCB CAD programs have such a steep learning curve.

Commercially made PCBs use plated through holes (PTH) to make connections between layers. Such boards would be very difficult to make using kitchen table technology. Most homebrew PCBs have a maximum of two copper layers. Home made double-sided boards use the component leads or short lengths of wire to make connections between the top and bottom copper layers.

Most PCB CAD software displays a top view of the PCB on the computer screen. This can be quite confusing when you are designing a conventional single-sided board. You are effectively drawing the bottom track layout while you are looking at the other side of the board. To avoid confusion, you can mark pin 1 of each IC or use IC pads which clearly identify pin 1. The standard DIL IC pads supplied with *PCB* (the software package) have a square pad for pin 1 and round pads for all other pins. This helps to avoid confusion. It is also a good idea to put some text on the board. Seen from above, text on the bottom copper layer will be a mirror image of the text that will be on the finished PCB.

PRINTING/PLOTTING. When the artwork is finished, we must find a way of getting the tracks and pads from the computer screen onto the copper clad PCB. There are several different ways of printing the etchresist pattern onto the board. Commercial PCB houses usually use a photographic process. The track pattern can be printed directly onto a transparent film using a plotter. Alternatively, the pattern can be printed on paper using a conventional plotter or printer. The pattern is then photographed with a special camera. The resulting photograph is used as a master artwork for exposing PCBs to ultra-violet (UV) light. The artwork is placed between special pre-coated photosensitive PCB and a UV light source. The photo-resist is softened by the UV light. When the board is developed, the exposed photo-resist will be washed away. Only the areas of photoresist that were shielded from the UV light by the opaque artwork will remain on the board.

Another method of producing PCBs uses silk-screen printing to print etch-resistant ink directly onto the copper board. The process of making the screen is quite similar to the photographic methods mentioned above. A fine mesh screen is coated with a UV sensitive emulsion. The screen is placed in close contact with the master artwork and exposed to UV light. The screen is developed by washing away the emulsion. The finished screen is used as a mask for spreading waterproof ink directly on the PCB surface. A rubber spreader called a squeegee is used to spread the ink. A single screen can be used to produce many PCBs in one session.

One of the most popular methods of homebrew PCB production uses toner from a laser printer or photocopier as the etch resist coating. The track pattern is printed onto a special transfer film using a standard laser printer or photocopier. The toner is then ironed onto the copper surface of the PCB using an ordinary clothes iron. Most laser printer toners are very waterproof and make excellent etch-resist. I have also had some success with plotting waterproof ink directly onto the PCB. I modified a Roland flat-bed plotter by removing the rubber pad from the plotter bed. As the rubber mat was approximately the same thickness as 1.6mm PCB, this left room to place the PCB directly on the plotter bed. I made quite a few boards using this method, some very successful, others not so good.

We will start by describing the process for making a PCB by the toner transfer method. The bottom copper layer from Figure 1.1 was used as the artwork. Remember that you will probably need to print a mirror image of your PCB layout on the transfer film. The process of ironing this image onto the board will mirror the image again and restore the original layout. Most CAD and graphics programs will have a mirror or flip function. Do a test print on plain paper before you commit yourself to using a sheet of expensive transfer paper. Any errors that you fail to spot at this stage could easily end up on the completed PCB! I know because I have made that mistake. Unfortunately, special PCB transfer films are far too expensive for the EI9GQ homebrew budget. A cheaper alternative will have to be found.

DISCLAIMER. Putting anything other than the manufacturer's recommended paper or film through your laser printer or photocopier could damage it. The printer/copier fuser unit is a high temperature roller used to melt the toner into the surface of photocopy paper. Some types of plastic OHP film or plastic coated paper could make a horrible mess of it. If in doubt, you can check the heat resistance of your paper by running a hot clothes iron over it. If you have any reservations about putting inkjet paper through your laser printer, then you should buy the 'proper' transfer film instead. I have had good results with TEC200 film. The most popular type of film is Press-N-Peel Blue which is available from Maplin and many other suppliers. See part 3 of G3XJP's excellent DSP transceiver project **[5]** for a good example of what can be achieved using toner transfer film. I have found that glossy photo paper for inkjet printers makes a very good toner transfer film **[6]**. Epson glossy photo paper gave particularly good results. I have also had good results from Xerox brand glossy photo paper.

Whichever brand of inkjet photo paper you decide to use, I would advise you to do a hot iron test to make sure it can survive the heat of the laser printer or copier.

TONER TRANSFER. If you are using Press-N-Peel or similar transfer film, you should ignore some of the advice below and follow the manufacturer's instructions. I printed the track pattern onto a sheet of Epson glossy photo paper. My artwork can be downloaded from

http://homepage.eircom.net/~ei9gq/art.html

Note that this track pattern has already been reversed so that is can be printed without modification. If you don't have a laser printer, the pattern can be printed on ordinary paper using an inkjet printer and photocopied on to the transfer film. The PCB should be thoroughly cleaned with fine steel wool and dried with a lint-free cloth. The photo paper or transfer film is then placed toner-side down on the surface of the copper PCB. One edge of the film can be secured with some paper masking tape.

The toner is then ironed onto the surface of the board using a hot clothes iron. The amount of time and pressure required will depend on the temperature of the iron and the type of toner used in the printer or copier. I use a hot iron at close to maximum setting. The iron is placed on the back of the photo paper and pressed firmly down for about 10 seconds while the toner sticks to the PCB surface. The iron is then moved slowly back and forth for a further 15 to 20 seconds. The amount of pressure required will depend to some extent on the complexity and size of the PCB pattern. Heavy pressure on the iron will tend to flatten the toner which will make lines slightly thicker and pads slightly larger.

Do not attempt to peel the photo paper off the PCB at this stage. The board should be placed in warm soapy water which will soften the photo paper. You should then unplug the iron and leave the room immediately. All homebrew radio enthusiasts have a natural curiosity and a tendency to poke at things, even when we know we shouldn't. Leave the board in the warm water for at least 20 minutes before removing the photo paper. By this time, the paper will be very wet and should be easy to remove. Any remaining paper residue can be brushed off with your fingertip while holding the board under running water.

Very stubborn residue can be removed with a coarse sponge. Don't be too afraid of damaging the track pattern, the toner is usually quite durable. Any small gaps in the toner can be repaired with an etchresist pen. **Photo 1.5** was the board after ironing.

ETCHING. The board was etched in a ferric chloride solution. Ferric chloride can be purchased in liquid or crystal form from most electronics suppliers. See the supplier's data-sheet for information about safe handling and proper disposal of ferric chloride. Ferric chloride is corrosive and quite toxic in concentrated form. However, it is probably no more dangerous than some household chemicals like bleach or drain cleaner. Wear rubber gloves and eye protection. Wear old clothes, ferric chloride stains won't wash out. Sodium persulphate is a good alternative etchant.

Etching time depends on the temperature, strength and condition of the etchant, the thickness of the copper foil and the area of copper to be removed. Commercial etching tanks use heated etchant and a stream of air bubbles or mechanical agitation to speed up the etching process. Boards that are upside down (with the copper facing downwards) will etch faster than boards that are lying copper side up. Double sided boards can be etched standing on edge if the etchant is deep enough.

Typical etching times range from five minutes for a heated/agitated tank up to about an hour for an unheated etching tank on a cold day. **Photo 1.6** shows the PCB in the etchant. Note that the PCB is floating face-down on top of the etchant. Also note the length of insulated wire that allows for

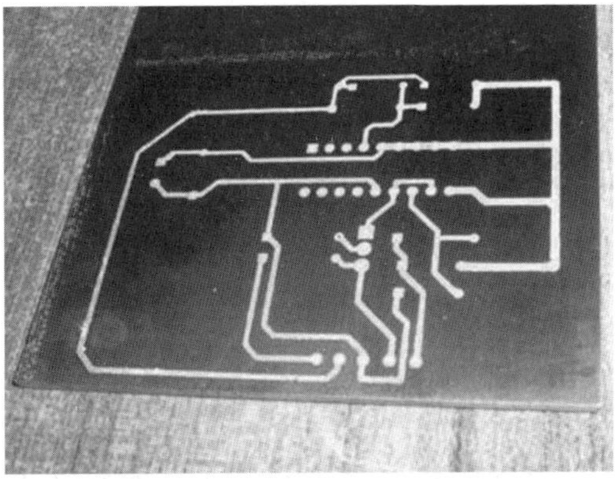

PHOTO 1.5: A PCB board after the artwork was imprinted on it using standard photo paper and an iron.

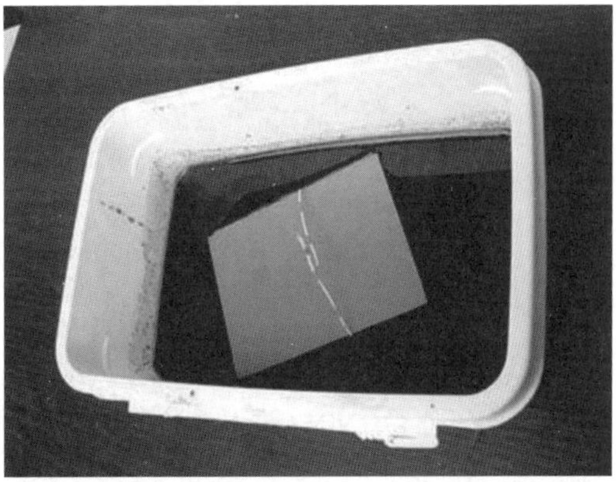

PHOTO 1.6: PCB in the etchant. Note that the etchant is both corrosive and toxic.

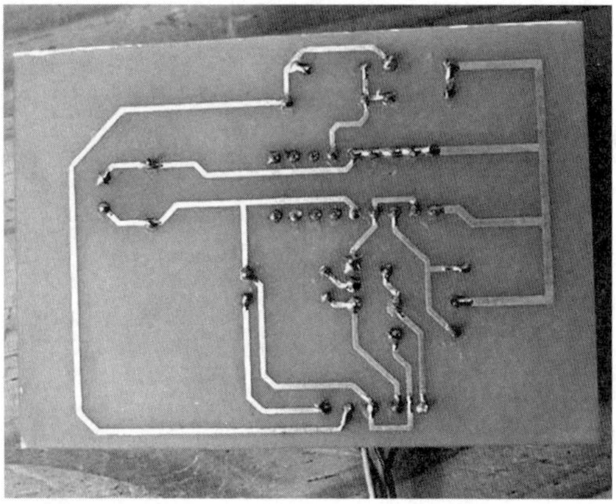

PHOTO 1.7. The bottom of our PCB showing the copper track and solder.

PHOTO 1.8: The top of our PCB showing the components.

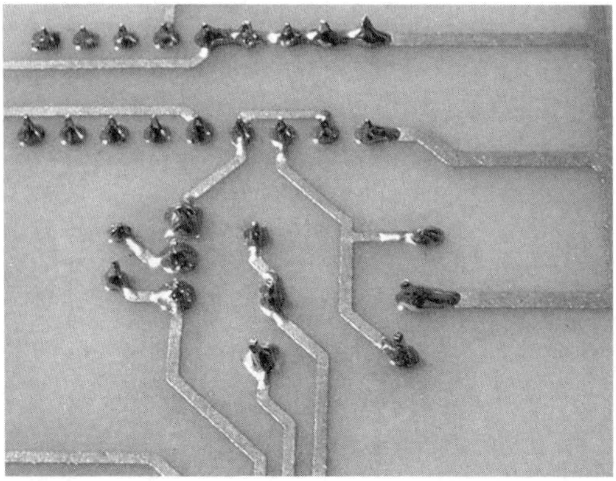

PHOTO 1.9: A close up of the PCB showing the high quality that can be achieved using cheap and simple techniques.

handling of the board without coming into contact with the etchant. The board should be wiggled from side to side several times during the etching process. This board took about 35 minutes to etch. I would recommend that you use a deep plastic tub rather than the shallow trays that are supplied with etching kits.

Once the board is etched it should be rinsed with clean water. If you find that there are still a few traces of unetched copper, you can put the board back in the etchant for a few minutes before you rinse it again. After etching, the toner can be removed from the board with paint/varnish remover or steel wool.

DRILLING. Unlike hand-drawn PCBs, CAD-produced PCBs tend to have very small pads for component leads. I used a 0.75mm drill to make the holes in this PCB. 0.8mm is about the largest drill size that can be used with this PCB layout. I normally use HSS (high speed steel) drills for this kind of work. Tungsten carbide drills take much longer to wear out but they break easily.

ASSEMBLY. The board should be cleaned again with fine steel wool. If you want to make a posh looking PCB, you can print the top silk-screen layer onto your photo paper or transfer film and iron it onto the top of the PCB. Next the board should be stuffed with components and soldered. I use a small brass-bristled spark-plug cleaning brush to remove any residual flux. As well as removing the flux, this takes the shine off the solder joints. If I want to make a really posh job of it, I re-flow all of the solder joints with the tip of the soldering iron to restore the shine.

As this is a well tried and tested circuit and it is unlikely to need repair or modification, I gave the copper side of the board a couple of coats of clear varnish. This should keep it in mint condition for many years. The varnish is plasti-kote clear enamel from B&Q. This method of PCB production produces boards that are not quite as good as the photographically produced boards from commercial PCB makers, but as you can see from **Photo 1.9**, they come pretty close. Note that the board in Photo 1.9 is not the result of a long trial and error process. You can expect to make boards as good as or better than this on your kitchen table. The PCB is for a simple PIC programmer.

PIC PROGRAMMER COMPONENT LIST			
C1	100nF	D2	13V Zener
C2	100μF	D3	5.1V Zener
R1, R2, R4	10kΩ	Q1	BC547
R3	1k5	18 pin IC socket (preferably ZIF)	
D1, D4, D5	1N4148		

Producing a PCB using photographic methods is a homebrew version of the system employed by most commercial PCB manufacturers. The circuit is a simple microphone amplifier for an SSB transmitter. The schematic is shown in **Figure 1.3**.

ARTWORK. As we will be using a pre-coated PCB with a positive photoresist, we will need a positive master artwork (copper areas opaque). Note that some PCB systems use a negative photo resist and require a negative master artwork (copper areas transparent). The artwork was produced

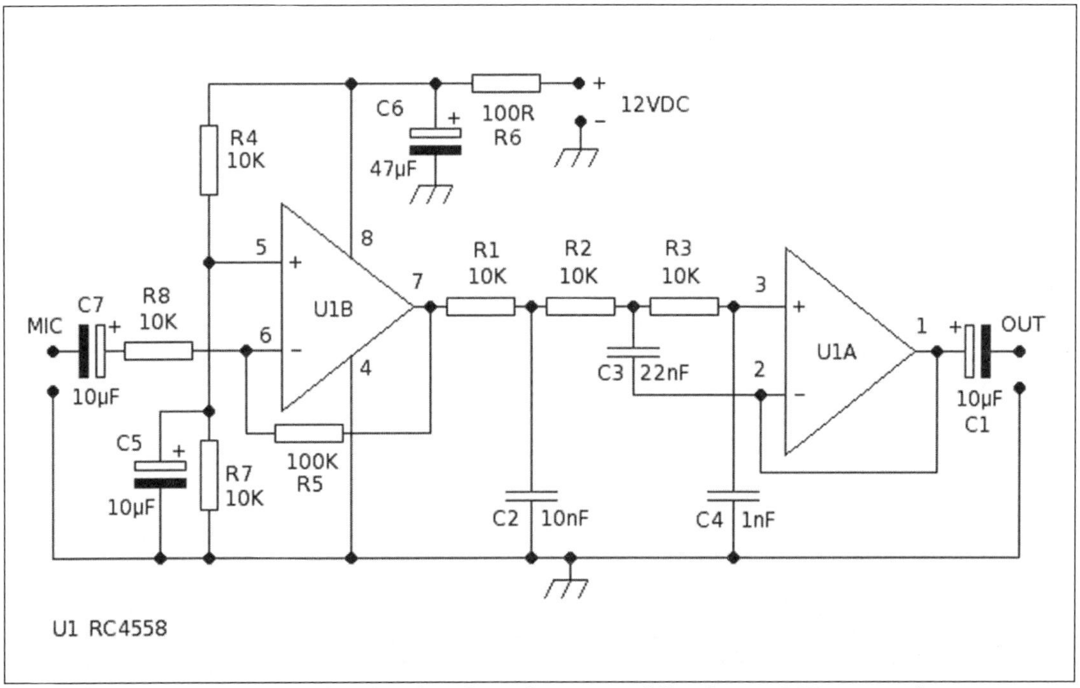

FIGURE 1.3: A schematic of a simple microphone amplifier for an SSB transmitter.

using PCB [7] CAD software. **Figure 1.4** shows the layout of the board. The copper track layer was printed on overhead projection film using a laser printer. If a laser printer is not available, the track pattern can be printed on white paper using an inkjet printer and photocopied onto the OHP film. Make sure you use the type of OHP film that is suitable for use in photocopiers and laser printers. The image should be printed on the side of the OHP film that is closest to the PCB. If the image is on the opposite side of the film, some UV light will creep around the edges of the opaque areas of the artwork. This will lead to thinner tracks and fuzzy edges. Do a test print on ordinary A4 copy paper before you print on the film. Hold the test print up to the light and look at it from the other side to check that the image is correct. You may find that you need to print a mirror image of the track pattern. Many High Street stationers will be able to photocopy onto OHP film. Remember to take an eight-pin IC socket with you so that you can check the size of the printed image.

High quality master artwork is the key to success. If the black areas are absolutely opaque, your PCB tracks will have sharp, clear edges and will be free of cracks and holes. You should set the printer/copier for maximum toner density. Contrast can be improved by stacking two sheets of film. This requires very careful alignment of the two sheets. The master artwork for this project was made from two sheets of laser OHP film. The track layout was printed using a 1,200 dpi HP laser printer with the toner density set to maximum. After printing, I cut off most of the surplus film around the printed area. One sheet was cut larger than the other so that I could use a strip of Sellotape to keep the sheets aligned. The alignment marks produced by most PCB software will help you to achieve perfect alignment. My artwork can be downloaded from

http://homepage.eircom.net/~ei9gq/art.html

EXPOSURE. Photosensitive PCB is coated with a UV-sensitive photoresist. Pre-coated PCB is available from most component suppliers. Alternatively, ordinary PCB can be coated with a photo sensitive layer (see hints and tips section). The artwork must be firmly held between the surface of the photo sensitive board and a sheet of clear glass. I used a sheet of glass from a picture frame and a simple clamp that was made from a strip of aluminium.

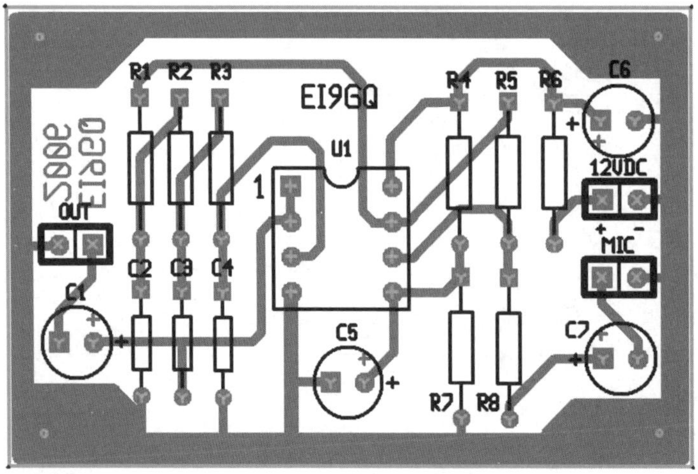

FIGURE 1.4: PCB layout for the microphone amplifier.

The artwork should be secured to the glass sheet with a single strip of Sellotape along one edge only. Make sure that the artwork is flat against the glass. If you are using double sheets, check that both sheets are properly aligned. The photo sensitive PCB should only be handled in subdued light. Once the protective film has been removed from the board, it should not be exposed to sunlight or bright fluorescent light. The board should be placed in contact with the toner side of the artwork and clamped firmly in place. **Photo 1.10** shows my home made exposure frame.

PHOTO 1.10: The homemade exposure frame.

After checking the alignment of the PCB and artwork one last time, the PCB should be exposed to a UV light source. I have a home-made UV light box that I use for exposing PCBs. Since most *RadCom* readers will not have a UV light box, I decided to try some of the alternatives.

The board can be exposed by placing the exposure frame in direct sunlight. The optimum exposure time will depend on the brightness of the sun, the thickness of the glass and the type of photoresist on the PCB. My first attempt was a complete failure. The board was overexposed and the tracks almost completely disappeared when the board was developed. The second attempt was much better. I exposed the board to direct sunlight for five minutes. The board was still very slightly overexposed but the end result was an almost perfect PCB. The best results were obtained with an exposure time of just two minutes.

After reading a few reports from successful PCB makers who use quartz-halogen lamps as a UV light source, I bought a 500W halogen light and mounted it on a shelf. The PCB exposure frame was placed a safe distance away from the front of the lamp. At a distance of 28cm from the glass cover of the lamp, the exposure frame gets quite warm but not too hot. For the first attempt, I used an exposure time of eight minutes. To my great surprise and delight, this produced a perfect PCB. The board needed almost five minutes in the developer before the unwanted etch resist was washed away. This is a little longer than normal, which would indicate that the board was slightly underexposed. For the next attempt, I increased the exposure time to ten minutes. Once again, the PCB was absolutely flawless and the developing time was a more reasonable three to four minutes. I have exposed several boards using the halogen lamp and so far I have enjoyed a 100% success rate.

Ultra violet fluorescent tubes of the kind used for disco lighting (black light), banknote scanners, suntan lamps or insect zappers can be used for PCB exposure. Ordinary fluorescent tubes can also be used, but exposure times will be much longer. A single Philips 30W TLD 30W/33 tube gave good results with an exposure time of one hour at a distance of 10cm. DON'T use EPROM eraser or germicidal UV tubes, because they produce short wavelength UV that is harmful to eyes and skin.

DEVELOPING. The two most commonly used PCB developers are sodium hydroxide and sodium metasilicate. Sodium hydroxide is widely available from most hardware stores as caustic soda or drain cleaner. Both types of developer can give excellent results. I have found that caustic soda is a bit more difficult to use. To get good results, the strength and temperature of the caustic solution must be carefully controlled. Sodium metasilicate-based developer is much less critical and will deliver consistent results over a wider temperature range. The developer used for the microphone amplifier boards was made by dissolving one teaspoon (5gm) of sodium hydroxide crystals (B&Q own brand caustic soda) in 500ml of water. You must resist the urge to add a little bit extra for good measure. This solution makes a good PCB developer at a typical room temperature of 17-20°C. Sodium metasilicate developer is available in liquid or crystal form from most electronic component suppliers. You should follow the supplier's instructions for mixing and safe handling.

The boards were developed in a plastic margarine tub (see **Photo 1.11**). The developer should be rocked very gently from side to side while the etch resist is washed away. You should wear eye protection while you are developing the board. You really don't want to get caustic chemicals in your eyes. After about one minute in the developer, the outline of the PCB tracks appears on the board. After two minutes, you should start to see bare copper. At three minutes, the board should be almost completely developed. The board should be left in the developer for a further minute to ensure that all of the exposed etch resist is washed away. The board should now be removed from the developer and thoroughly rinsed with running water. If your developing time is longer than 4-5 minutes, you probably need to increase the UV exposure time or slightly increase the strength of the developing solution. The remaining photo resist should be an exact replica of the master artwork. The board

PHOTO 1.11: The margarine tub used for developing boards.

PHOTO 1.12: Etching tub. Note the wire aro-und the board for safe handling of the board.

should be handled very carefully at this stage. The resist must not be scratched before the board is etched.

ETCHING. See previous section for details of how to etch a PCB using ferric chloride as an etchant. I have always used ferric chloride as a PCB etchant in the past, so I have no experience of using any other etchant. Just for the sake of trying something different, I used a sodium persulphate etchant for this project. Sodium persulphate is available from Maplin and many other suppliers. I mixed 100g of sodium persulphate with 500ml of warm water as specified in the supplier's instructions. When the powder was fully dissolved, I placed the PCB in the warm etchant. **Photo 1.12** shows the etching process. Note the insulated wire which allows for handling of the board without coming into contact with the etchant. The board was fully etched in 13 minutes. Sodium persulphate works best when it is warm. A heat lamp can be used to warm the etchant from above. I used a fridge thermometer to monitor the temperature of the etchant and the halogen lamp to provide the heat. Be very careful about how you do this. A melted container of etchant would make a really horrible mess. A glass container would be safer than my plastic tub. You need to stay close enough to monitor the temperature, but you should avoid breathing the vapour from the warm etchant. The board should be wiggled gently from side to side several times during the etching process. This helps to clear etched copper from the surface of the board.

After etching, the board should be rinsed thoroughly under running water. A quick splash under the tap is definitely not good enough. When you are ready to drill the board and populate it with components, the etch resist can be stripped off the board with paint/varnish remover or very fine steel wool. Alternatively, the board can be re-exposed to your UV light source without using the artwork as a mask or left in bright sunlight for a few minutes. The etch resist can then be removed by putting the board back in the developer for a few minutes. If you are not planning to use the board for a while, the etch resist can be left on the board as a protective coating.

DRILLING. Hand-cranked egg-beater style drills are just about usable for hand-drawn PCBs. To drill the tiny holes required by CAD produced PCB layouts, however, a 12V miniature PCB drill or a mains powered rotary tool will do a much better job.

After many years of breaking drill bits, I have at last discovered a reliable way of drilling holes in PCBs. I use a 12V PCB drill with a reduced supply voltage of around 6 volts. Using a lower supply voltage reduces the kick caused by the torque reaction when the drill starts turning. The lower voltage also reduces the speed of the drill, which results in longer life from HSS drill bits. I use a push-to-make foot switch to operate the drill. This leaves both hands free to hold the drill and PCB. You should wear safety goggles. If you don't have any, go and get some now. To avoid drilling holes in the table, you should drill the board on an old telephone book or a paperback.

The exact procedure I use is as follows: Place the tip of the drill in the centre of the pad to be drilled. I use a bright light so that I can see what I'm doing. Taking care to keep the drill absolutely vertical, press the foot switch.

PHOTO 1.13: The PCB board after drilling.

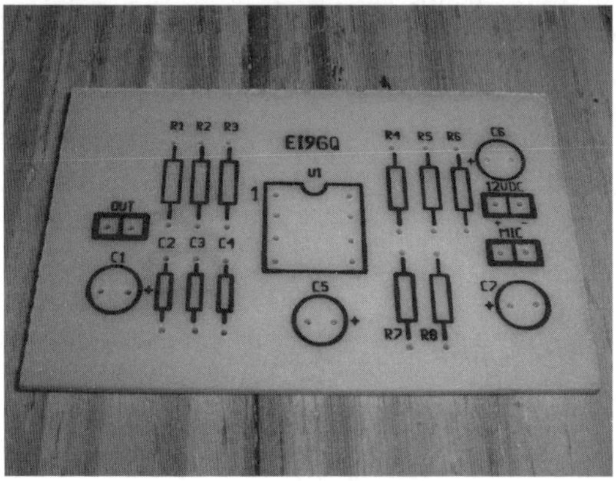

PHOTO 1.14: The silk-screen layer from the cad software printed on the board.

Don't hold the drill too firmly. The hole in the centre of the pad will act as a guide for the drill. Even if you are a little off centre, the drill will tend to centre itself with the hole in the copper pad. When the hole has been drilled, withdraw the drill vertically while it is still turning and release the foot switch. I used a 0.75mm drill for this PCB. **Photo 1.13** shows the drilled board.

Most constructors recommend using a drill stand for PCB work. I prefer to use a hand-held drill and the above technique. I have drilled several hundred holes during the last couple of weeks without breaking a single drill bit or drilling an off-centre hole.

Because most of my PCBs end up in a screened box with the lid screwed on, I don't really care how they look. This time I printed the top silk-screen layer from the CAD software on the top of the board (see **Photo 1.14**). The pattern was printed on glossy photo paper and ironed onto the top of the board. The same warnings apply to this process as I detailed previously.

DOUBLE-SIDED BOARDS. There are a few different ways of making double sided boards using photographic methods. It is possible to process both sides of the board at the same time by using a double sided exposure frame and a pair of UV light sources. This requires extremely accurate alignment of the artwork on both sides. I find it is a lot easier to do one side at a time. The first side of the board is exposed, developed, etched and drilled as it would be for a single sided board. Use a sharp new drill to reduce the risk of damaging the photoresist and protective coating on the opposite side. When the first side is finished, it can be covered with strips of PVC tape to protect it while you are etching the second side. Now you can remove the plastic protective film from the second side of the board. The artwork for the second side should be aligned carefully with the holes in the board. This ensures that both sides of the board are properly aligned. The second side of the board can now be exposed, developed and etched. Be very careful to protect the first side from the etchant while you are etching the second side. After etching, the protective tape and photoresist can be removed, hopefully to reveal a perfect double-sided PCB.

If you are really paranoid about protecting the first side of the board, you can give it a coat of paint or varnish before you start work on the second side. The paint can be removed with paint/varnish remover after the second side is etched. I have found that using a few strips of PVC insulating tape is just as effective.

SURFACE MOUNT. SMT (surface mount technology) boards as shown in **Photo 1.15** are made using exactly the same methods as boards for leaded components. PCBs with leaded components are often referred to as PTH (plated through hole) technology. This is a bit of a misnomer in our case because most amateurs will not have the equipment to make boards with electroplated holes. I have used the photographic methods described here to make boards using surface-mount ICs and 603-type chip capacitors and resistors. The main differences between boards for leaded components and boards for SMT are the methods used for mounting and soldering the components.

PHOTO 1.15: here is the same microphone amplifier circuit built using surface mount technology (SMT).

HINTS AND TIPS. As an alternative to pre-coated PCB, you can apply a coat of photo resist to ordinary PCB laminate. The photo resist liquid can be sprayed or wiped onto the surface of the PCB. Extreme cleanliness is required for this process. See Harry's Homebrew Homepage **[8]** for a good tutorial.

Double-sided boards can also be made by etching a pair of single-sided boards and sticking them together with epoxy resin. If you use 0.8mm laminate, the finished product won't be much thicker than standard 1.6mm board.

For small circuit boards, you can get more mileage per sheet of film by printing two copies of your layout on each sheet. I find this useful because I usually use a double layer of film for the artwork. This way I get two for the price of one.

Large unused areas on your PCB should be filled with grounded copper. Apart from the electrical advantage of having a better ground plane, you will use a lot less etchant and developer.

Not all PCBs are created by chemical etching. Some PCB prototyping systems use a cutting tool to remove unwanted copper from the board. At least one of the regulars on the 80m homebrew net (3,727kHz) makes all of his PCBs with a home-made PCB cutting/milling machine.

Once again, I will remind you to wear eye protection. Remember the old saying "It's all fun and games until someone loses an eye".

READERS' HINTS AND TIPS. Many thanks to the readers who have sent me their comments and suggestions. The PCB-making techniques previously covered in *RadCom* prompted some readers to share their experiences.

James, G3RUH, describes his method of making PCBs, "Another effective method is to use 'PCB Transfers'. See RS part number RS 436-083. These are press-on shapes reminiscent of Letraset.

1. Print your artwork at 1:1 onto paper and check with a ruler that it's correctly sized!
2. Sellotape the artwork to the copper-clad board.
3. Use a fine punch to dot all the holes through the paper artwork.

4. Now use the PCB transfers to draw the artwork onto the PCB. Do the ICs first, then the component pads, and then the tracks. Regularly burnish the tracks etc using a finger nail, with a piece of the semi-transparent protective interleave sheets as a buffer. Touch up with etch-resistant pen.
5. Etch.
6. Remove artwork with wire wool etc."

Ted, G3NPL, reports that he gets excellent results when he uses his inkjet printer to print PCB artwork onto overhead projection film. I haven't enjoyed much success with my own old Canon i350 inkjet. Obviously, the newer photographic printers are better suited to the job. Ted writes: "I have been using an inkjet of the newer type for printing photographs, in this case a CANON PIXMA iP2000. It gives excellent results on all boards. It must be set to high resolution and with overhead film set to high resolution paper."

Alan, G4GQL, sends this gem of a tip for improving the contrast of laser printed or photocopied artwork. Alan writes: "Lay the positive on a flat table over a piece of paper bigger than the positive. With a piece of cotton wool dipped in white spirit carefully wipe over the image - it will turn jet black. Try it and hold it up to the light - you will be amazed how dense it is. You must use a good white spirit. And dry the film in front of a fan heater."

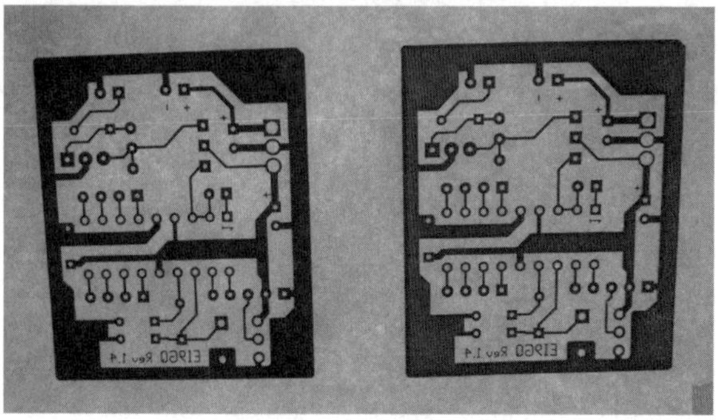

PHOTO 1.16: Homebrew reader Alan, G4GQL uses a clever technique to improve the contrast of laser-printed or photocopied artwork.

I tried this technique and, as Alan promised, I was amazed at the result. **Photo 1.16** shows an OHP transparency with two identical PCB track patterns printed on it. I won't tell you which side has been treated with white spirit. Alan recommends a fan heater for drying the film. I used a 1,400W hair-dryer. Please be careful when you are drying the film. A powerful heater and a bottle of white spirit could be a very dangerous combination.

EQUIPMENT USED

Small photograph frame (for the glass)
500W halogen lamp (B&Q)
HP LaserJet 1200 printer
Pre-coated photo sensitive PCB Maplin JP56
Staedtler Lumocolor transparent photocopy/laser film (NOT inkjet type)
Sodium persulphate etchant 100g Maplin NO6CG
Sodium hydroxide. (B&Q caustic soda)
Photographic or fridge thermometer
12V PCB mini-drill
Assorted plastic tubs and a roll of Sellotape

MICROPHONE AMPLIFIER COMPONENT LIST			
R1, R2, R3,	10kW	C3	22nF
R4, R7, R8		C4	1nF
R5	100kW	C6	47µF electrolytic
R6	100W	U1	RC4558 or
C1, C5, C7	10µF electrolytic		similar dual op-amp
C2	10nF		(TL072, LF353)

Metal Bashing

A lot of my radio projects are housed in commercially made metal boxes. Many component suppliers have a good range of enclosures available at a reasonable price. Aluminium sheet has become so expensive that it is often cheaper to buy a simple aluminium box than it is to make one. The more rigid die-cast alloy boxes are quite expensive, but they are ideal for projects where mechanical stability and good screening are important. For larger projects, enclosures that have a vinyl coated steel lid and aluminium bottom half are easy to work with and very good value. It is often possible to re-use enclosures from scrap equipment to house your projects. The steel enclosure used for the power supply project was recovered from a consignment of scrap medical equipment.

Heavy steel enclosures are quite difficult to work with. Cutting out rectangular or round holes in sheet steel can be quite a challenging task when you only have a few hand tools. I still have nightmares about the beautiful shiny stainless steel boxes I acquired from a local scrap yard. I couldn't cut them with a hacksaw or even drill holes in them!

When you find that none of the available enclosures is suitable for your project, it is often easier to make a new box from scratch than it is modify an existing one. I built the enclosure for my HF transceiver **[9]** because I couldn't find a commercially made one that was suitable for the job and there was nothing suitable in the scrap yard. I have also built special enclosures for VHF/UHF helical and cavity filters. The stripline enclosure of my VHF linear amp is made from a sheet of 1mm aluminium.

I have very little experience of working with sheet metal and very few tools. My current metalworking kit consists of a blunt hacksaw, a power drill with a few HSS bits (mostly blunt or broken), an electric jigsaw, a few simple hand tools, hammer, screwdrivers, pliers etc. A strong pair of scissors can be used to cut aluminium sheet up to a thickness of about 1mm. The two most useful tools in my metalworking kit are a sharp carpet knife and a small nibbling tool. The least useful tool is the jigsaw which is far too fast and powerful for delicate work.

SAFETY. Always wear safety goggles when working with power tools. Using tools for purposes other than the purpose they were designed for can be particularly dangerous.

Apart from the fact that you will be using a much stronger pair of scissors or shears, the process of making an enclosure is much the same as making a new suit of clothes. You will need to decide what size box will be required, make a pattern for the various panels, cut the panels from sheet metal and form them into the required shape. The design of the enclosure will depend on your own individual requirements. The enclosure can be just a simple box or a more elaborate structure like a chassis with circuits mounted on both sides. Many homebrew projects call for an enclosure which is divided into separate compartments to provide improved screening of the enclosed circuit.

When you have decided on a layout and size for your project, you should write down the exact measurements of your enclosure and spend some time drawing up a detailed pattern for the metal panels. The pattern can be drawn by hand on graph paper or with a computer CAD (computer aided design) or drawing program. I use a simple vector drawing program called Xfig on a Linux PC. Similar software is available for Windows and Mac OS. The pattern should be drawn to scale. Once you have a detailed pattern of the panels, you can cut them from sheet metal. I usually use 1mm aluminium sheet for my homebrew projects because it is relatively easy to cut and bend. I will assume that like me, you don't have access to proper metalworking tools like a guillotine or bending brake.

I have tried using an electric jigsaw for cutting sheet aluminium, but I found that I could never manage to cut a straight line. The jigsaw also tended to leave a slightly jagged edge on the cut. A simple 12in hacksaw is a bit slower, but it usually does a better job. There is a much better way of cutting straight edges in aluminium sheet. If you use a very sharp knife to cut a deep groove in the surface of the aluminium sheet, you will find that it is quite easy to bend the sheet along the line of the cut. The sheet can be bent around a sharp metal edge like a length of angle iron or the edge of a metal bench. Smaller pieces can be bent in a bench vice. Once the sheet has been folded back through 180 degrees, it is very easy to break off the bent section. This method gives a perfectly straight cut every time. A steel rule will help you to draw a straight line and a Stanley or similar type of carpet knife can be used to cut into the aluminium sheet. Be careful not to ap-

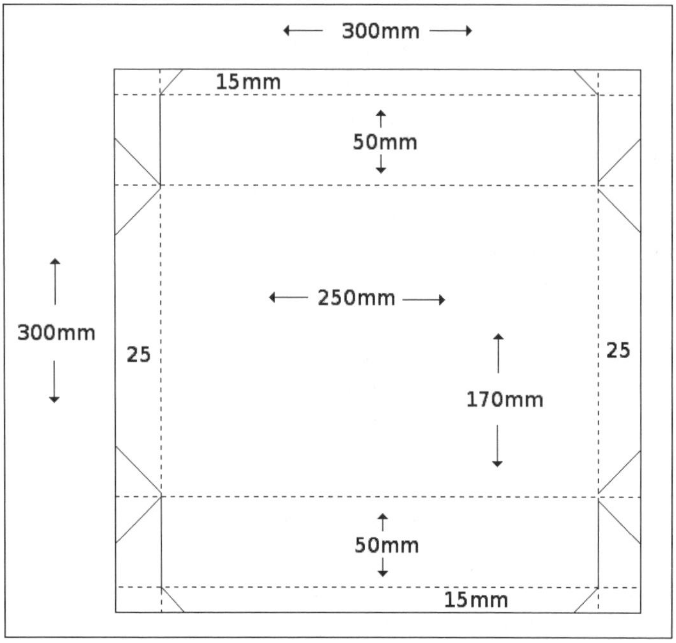

FIGURE 1.5: Template for the base of the box.

ply too much pressure to the carpet knife in case it breaks. A metal bodied carpet knife is much safer than plastic types. Retractable knives with plastic bodies and renewable blades with snap-off sections are not suitable for this job.

You can use a similar technique to form nice clean 90° bends in your aluminium sheet. If you use the carpet knife to score along the line of the bend, it will help you make a nice straight and sharp bend. The line should be scored on the outside radius of the bend. You should only use the knife on one side of the sheet. Don't cut too deep or the sheet might tend to break along the cut.

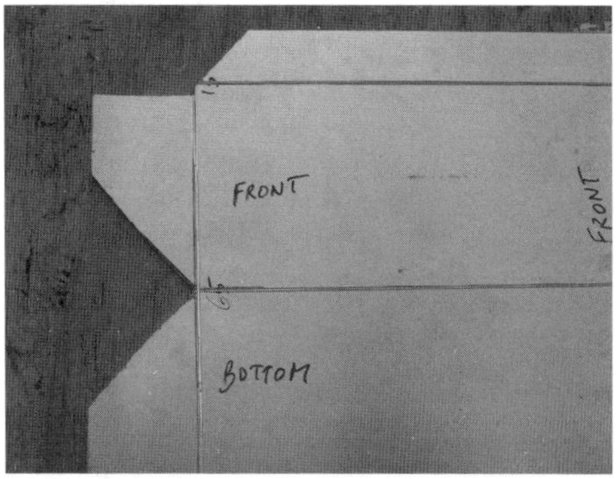

PHOTO 1.17: Close-up of the cuts and fold lines. Don't worry about slightly ragged edges – these will be covered by the lid.

Building a Transverter Enclosure

The transverter project **(page 131)** will be housed in a home-made box made from 1mm aluminium sheet. The only tools used for the construction of the enclosure were a sharp knife, a hacksaw, a steel rule, measuring tape, a bench vice and a felt tip pen. **Figure 1.5** shows the pattern for the bottom half of the box. A 300mm x 300mm square of 1mm aluminium was cut from a larger sheet using a carpet knife and straight edge. The dimensions of the bottom half of the case will be approximately 250mm wide, 170mm deep and 50mm high. The finished project will be slightly bigger than this because of the thickness of the sheet used for the top half of the enclosure and the way that the sheet is bent. The sheet should be carefully marked, scored and cut as shown in **Photo 1.17**. Remember the old carpenters maxim "measure twice, cut once".

PHOTO 1.18: The folded base panel for our box.

Once the panel has been scored and the corners are cut out, the box can be bent into shape. The larger sections are quite easy to bend. The 15mm lip at the top of the front and rear panel is more difficult to bend. This job is a lot easier if you use a bench vice. The top half of the project box is made from a single sheet of metal which is bent into a U shape. After it was bent into shape, the bottom half of my metal box was almost 252mm wide and 171mm deep. I used a 354mm x 172mm sheet of 1mm aluminium for the lid. **Photo 1.18** shows the bottom half of the box, while **Photo 1.19** shows the completed enclosure.

PHOTO 1.19: Adding the lid makes for a professional-looking enclosure.

You might prefer to use steel sheet or fancy perforated steel for the top half of the enclosure. Even if you use a proper etching primer, paint doesn't stick very well to aluminium. It is very easy to paint a steel cover with Hammerite or similar tough enamel.

References

[2] www.k7qo.net

[3] *VHF/UHF Manual* Chapter 3. G.R. Jessop, G6JP. RSGB 1983

[4] http://pcb.sourceforge.net

[5] PIC-A-STAR Part 3. *RadCom* October 2002.

[6] http://homepage.eircom.net/~ei9gq/pcb.html

[7] http://pcb.sourceforge.net

[8] http://web.telia.com/~u85920178/data/pcb2.htm

[9] http://tinyurl.com/ygnfdg

2

Building a Receiver

THE FIRST thing your home made amateur radio station will need is a receiver. We will begin by building a two stage audio amplifier. This can be used as the audio stages of an amateur band receiver. It can be used with a simple front-end to make a regenerative or a direct conversion receiver or as part of a more elaborate superhet receiver. It is also useful as a general purpose audio amp and signal tracer.

The first stage is a TL071CN **[10]** op-amp which is configured as an inverting amplifier with a gain of around 20dB. The output stage is an LM386N **[11]** power amplifier. The second stage has about 26dB gain and a maximum output of several hundred mW. Distortion is ~0.1% THD at 100mW output.

I built two versions of the audio amplifier, one on Veroboard (**Photo 2.1**) and the other Dead Bug style (**Photo 2.2**). Both work equally well. The Dead Bug version took a lot less time and effort to build. Because both stages of the amplifier are operated at relatively low gain, the physical layout of the circuit is not very critical. You can use any construction method you like. The components are all readily available. I bought the ICs and resistors from Maplin. The electrolytic capacitors are either 25V or 16V working voltage. The resistors are 0.25W or 0.5W metal film.

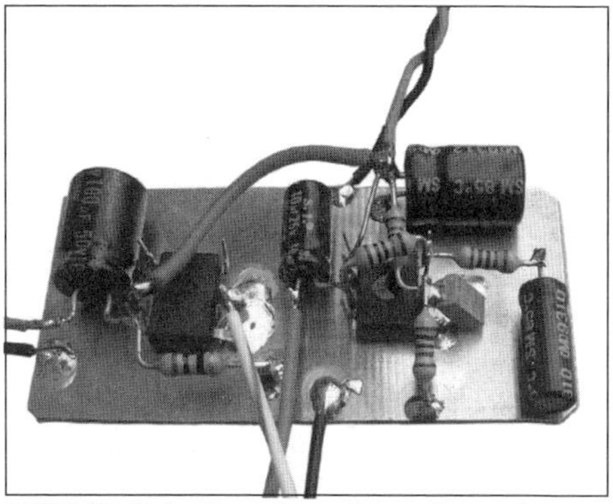

PHOTO 2.1: Veroboard version of the amplifier.

The amplifier can be powered by a 9V battery or a 12V DC supply. I have tested the amplifier at various voltages between 5V and 15V. Please note that the LM386 data sheet [11] specifies an absolute maximum supply voltage of 15V for the LM386N-1. The TL071CN op-amp is pin compatible with the ubiquitous uA741/LM741 and several other 8-pin op-amps. If you don't have a TL071 available, you can substitute a similar device like the LF351.

PHOTO 2.2: Dead-bug version of the amplifier, ugly but very effective.

COMPONENT LIST			
IC1	TL071CN	C1	1μF electrolytic.
IC2	LM386N-1	C2, C3	10μF electrolytic.
R1, R3, R4	10kΩ	C4	330μF electrolytic.
R2	100kΩ	C5	0.1μF (100nF) plastic or ceramic.
R5	10kΩ log.		
R6	10Ω	C6	100μF electrolytic
		All electrolytic capacitors 25V or 16V	

Direct Conversion (DC) Receiver Front-end and Local Oscillator

Next we will be building a direct conversion (DC) receiver front-end and local oscillator for operation on the 80m band. This receiver front-end can be used with the audio amplifier described previously to make a complete SSB/CW receiver.

The DC receiver is a very good first project for the homebrew radio enthusiast. **Figure 2.3** is a block diagram of the receiver. The first block is a band pass filter with a 1dB bandwidth of about 400kHz from 3.5 to 3.9MHz. The next block is a product detector (mixer) and an audio low pass filter. In the third block, the audio output from the product detector (PD) is amplified by a two stage AF amplifier. The output from the low level amp can be used to drive the audio power amplifier built previously or any similar audio amplifier. The final block is a 3.5 to 3.8MHz variable frequency oscillator (VFO) and buffer amplifier.

CONSTRUCTION. The band pass filter (BPF) was built using point-to-point wiring (dead bug) on some fibreglass PCB (see **Figure 2.4** and **Photo 2.3**). The three inductors L1, L2 and L3 were wound on Toko 10K type coil

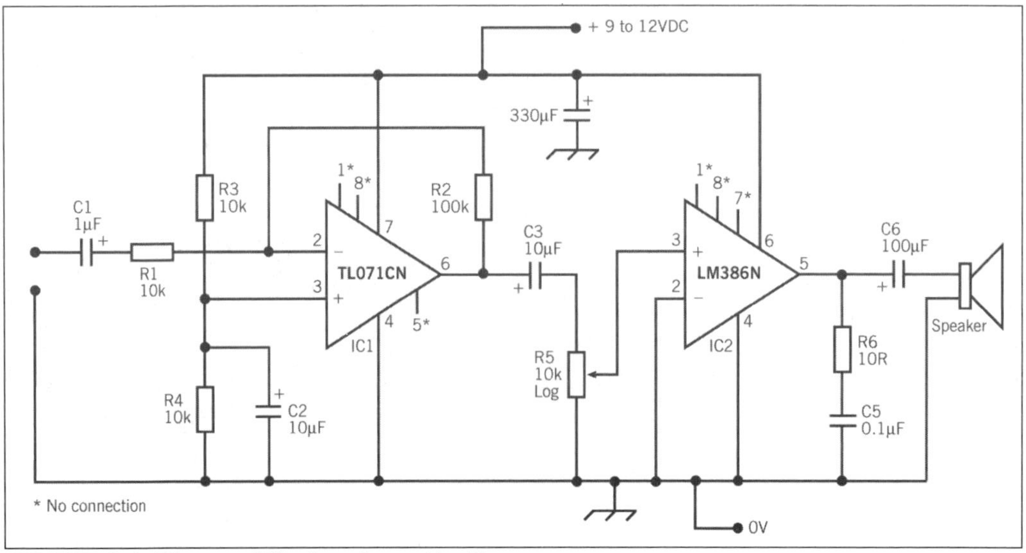

FIGURE 2.1: The amplifier's schematic diagram.

formers. Other types of slug tuned coil formers can also be used. To achieve the required inductance of 10µH will take about 25 turns on a Toko 10K former. All capacitors are standard value ceramic types.

As an alternative to using slug tuned inductors, you might prefer to use fixed value 10µH inductors for L1, L2 and L3. You can make suitable inductors by winding 45 turns of enamelled wire on a T50-2 powdered iron core. If you use fixed inductors in the BPF, you will have to use variable trimmer capacitors for C2, C4 and C6. As 150pF and 180pF capacitors are not easy to find, you can use a fixed 120pF capacitor in parallel with a 60pF trimmer for C4 and a 150pF fixed capacitor in parallel with a 60pF trimmer for C6 and C7. **Figure 2.5** shows a plot of the frequency response of the BPF.

The product detector (PD) is a simple balanced mixer using four 1N4148 or similar silicon diodes (See **Figure 2.6 (overleaf)** and **Photo 2.4**). My first prototype used an SBL1 packaged Schottky mixer module. After some experimenting, I found that the homebrew mixer was much less prone to breakthrough from AM broadcast stations. I live within a few hundred metres of the RTE 10kW Cork medium wave transmitter. The current version of the receiver seems completely immune

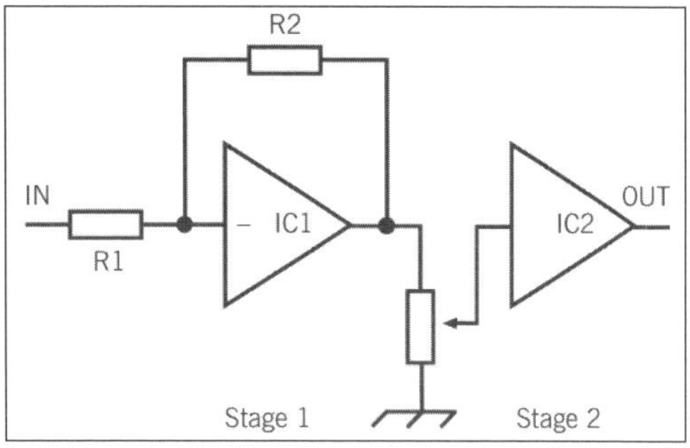

FIGURE 2.2: Simplified version of the amplifier.

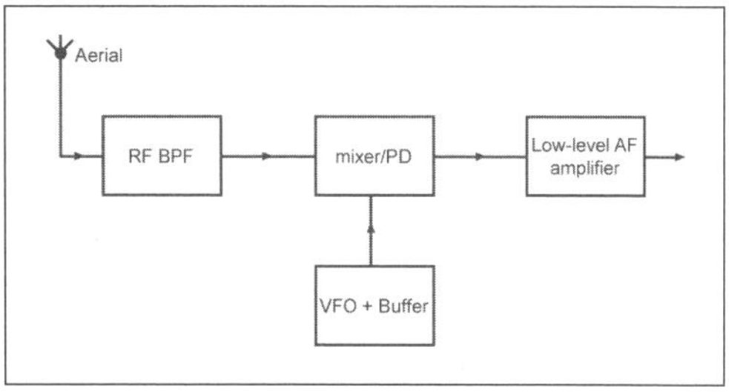

FIGURE 2.3: A block diagram of our receiver showing the four key modules.

PHOTO 2.3: The band pass filter was built using point-to-point wiring on glassfibre PCB.

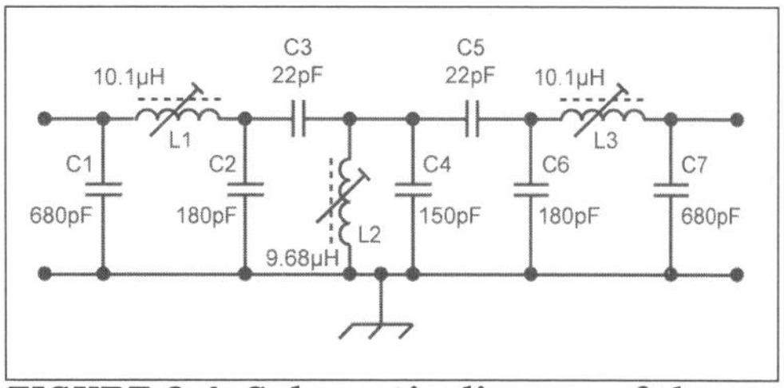

FIGURE 2.4: Schematic diagram of the band pass filter.

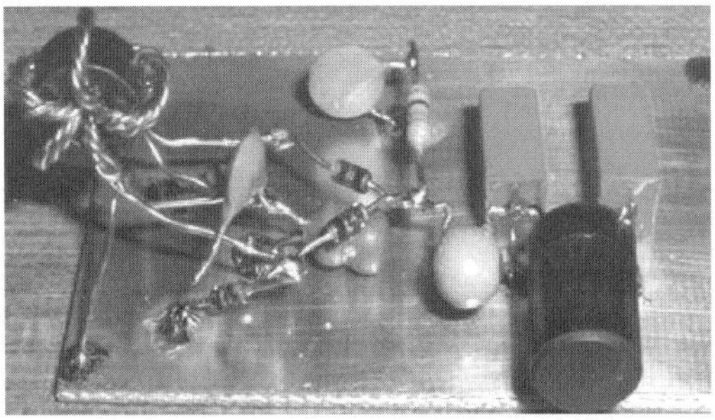

PHOTO 2.4. The product detector built using dead bug construction.

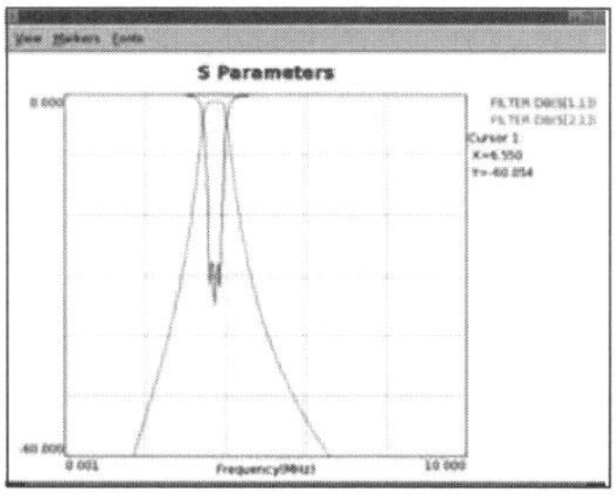

FIGURE 2.5: A plot of the frequency response of the band pass filter.

to breakthrough, even when listening to very weak DX stations in the early hours of the morning.

The PD stage was also built using dead bug style construction. T1 is a broadband transformer consisting of five-turns, trifilar wound on a ferrite toroid core. To make a trifilar winding, twist three equal lengths of insulated wire tightly together. Make sure that the three windings on T1 are connected properly. Take care to connect the diodes correctly. I got them wrong on my first attempt. You can use very thin PVC insulated wire or enamelled copper wire for the trifilar windings. R1 provides a 50Ω termination for the diode mixer at radio frequencies.

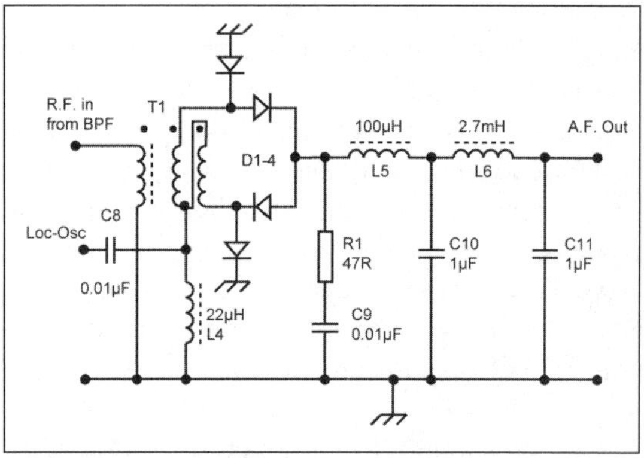

FIGURE 2.6: The product detector circuit diagram.

The local oscillator I used (see **Figure 2.7** and **Photo 2.5**) was one I had designed for an older project of mine. The oscillator and first buffer amplifier are BF245C junction FETs (JFETs). Other N-channel JFETs like the BF256 or 2N3819 should also be suitable. The second buffer amplifier is a PN2222 NPN transistor. The local oscillator was built on a 'proper' etched and drilled PCB. The track pattern was drawn by hand with an etch resist pen. A dead bug version should work just as well.

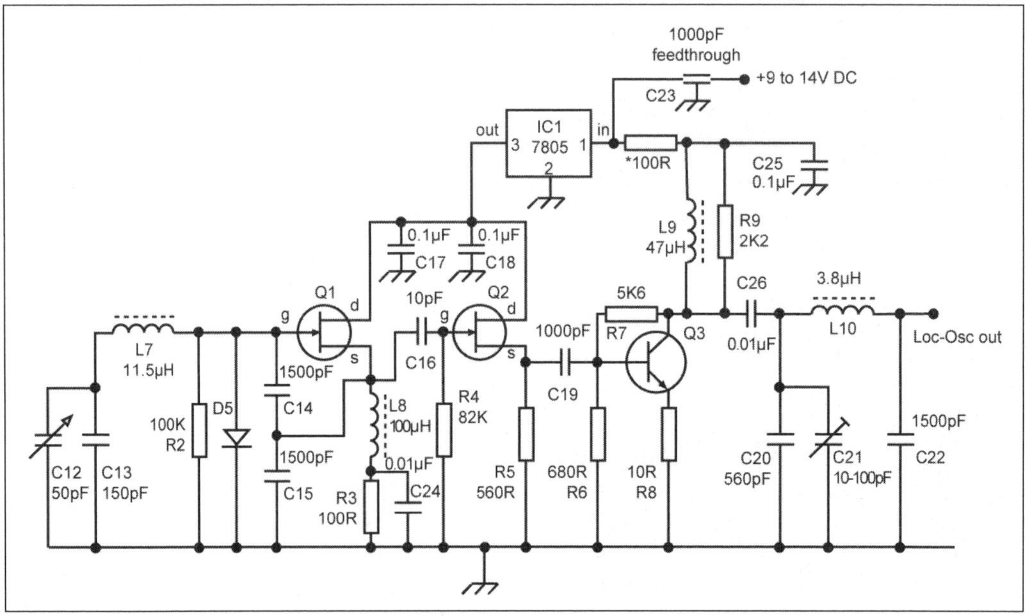

FIGURE 2.7: The variable frequency oscillator schematic diagram.

PHOTO 2.5: This VFO was built on 'proper' etched and drilled PCB.

Variable frequency oscillator (VFO) construction is often regarded as a black art. Of the many VFOs I have built over the years, some have worked very well, most have worked reasonably well and a few have been dreadfully poor. Some of the worst examples were carefully built in a solid metal enclosure using all of the recommended components. Some of the best VFOs were hastily thrown together using all the wrong components. I am not advocating sloppy construction as the key to success as a VFO builder. I am just trying to encourage you to experiment with different circuit layouts and methods of construction.

The recipe for a stable VFO is mechanical stability, thermal stability and good regulation of the VFO supply voltage. The VFO should be followed by one or more buffer amplifiers. Mechanical stability can be achieved by enclosing the VFO in a rigid metal box and making sure that all components are properly secured in position. A small dab of Araldite or similar epoxy resin can be used to keep components in place. Don't get too carried away with the epoxy, you might want to change a component later.

Thermal stability is one of the most important aspects of VFO design and construction. You should use capacitors with good thermal stability. Polystyrene, silvered mica and NPO type ceramics are all recommended. I have had the most success with polystyrene capacitors. If the VFO is tuned by a variable capacitor, the main tuning capacitor should be a high quality, air-spaced, double bearing type. The other critical component in a free running LC VFO is the inductor.

The best inductors will have a high unloaded Q and good thermal stability. The inductors used in many commercial VFOs are made from silver plated copper wire, wound on a ceramic former. Sometimes there has to be a compromise between size, cost and thermal stability. The low cost option is to use a small Toko or similar inductor with a ferrite core. This will fit in a small space but will have relatively poor thermal stability. Powdered iron cores usually have better thermal stability than ferrite. Powdered iron cores with known characteristics are made by Amidon, Micrometals and others.

I tested two different inductors in the VFO of this receiver. The best one had 60 turns of 0.5mm copper wire wound on a ceramic former. The ceramic former is a section of the ceramic element from a 1kW electric heater. Don't try to drill holes in this material. I destroyed several HSS drill bits before I gave up.

For the second attempt, I used 48 turns of much thinner wire wound on a T50-2 powdered iron core. Frequency stability of the VFO was very good with the larger ceramic cored coil. The VFO drifted a couple of hundred Hz in the first few minutes and remained stable to within ±100Hz per hour thereafter. Frequency stability with the powdered iron core was not quite as good, but still reasonable at ± a few hundred Hz per hour.

The low level audio amplifier (See **Figure 2.8** and **Photo 2.6**) uses a pair of transistors in a direct coupled circuit. The first stage is an NPN transistor in a common base configuration. The second stage is a PNP transistor in the more usual common emitter configuration. I used Japanese 2SC945 and 2SA733 transistors. Any general purpose transistors like the BC548 NPN and BC558 PNP should work just as well.

ALIGNMENT AND TESTING. Each of the stages in the block diagram (Figure 2.3) was built as a separate module. There are many advantages to this modular approach.

* Each module can be built as a separate project.
* Individual stages can easily be upgraded or repaired without disturbing the other building blocks in the system.
* Unused blocks can be reused in later projects.

PHOTO 2.6: The low level audio amplifier.

You will probably find that aligning and testing the VFO are the most difficult parts of this project. Your task will be a lot easier if you have a frequency counter or a HF receiver with a calibrated frequency display. The choice of a fixed inductor for L7 means that you may have to experiment with the value of C13 to get full coverage of the 80m band. If you find that your VFO tunes from 3.3 to 3.7MHz instead of tuning from just below 3.5MHz to just above 3.8MHz, you will need to decrease the value of C13. If the VFO frequency is too high, you will need to increase the value of C13. Alternatively, you can adjust the number of turns on L7 until you get the required coverage.

Once you have got the VFO running at the correct frequency, put a 47Ω or 51Ω resistor across the output of the final buffer amplifier to act as a dummy load and tune the VFO to the middle of the 80m band (3.65MHz). Then adjust C21 for maximum output. You can measure the VFO output with an oscilloscope or RF voltmeter. If you don't have such an instrument, you can

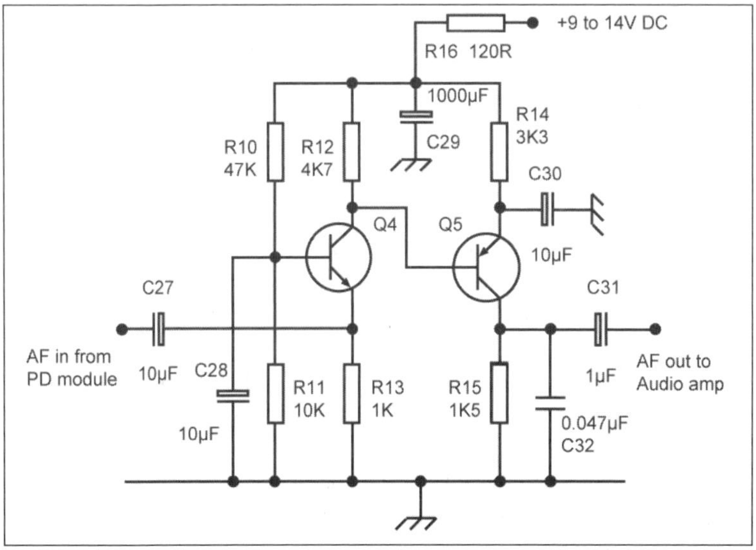

FIGURE 2.8: Schematic diagram of the low level audio amplifier.

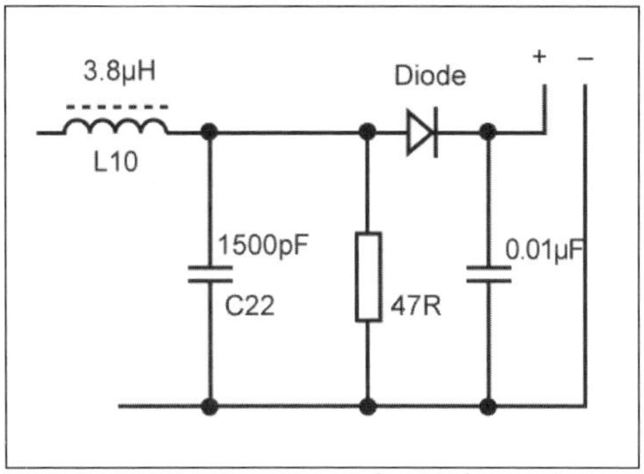

FIGURE 2.9: An RF probe. Why not try building one yourself?

use the simple RF probe in **Figure 2.9**. When you have finished with the alignment of the VFO, remove the extra components and put them on the shelf for next time. A simple diode probe is one of the most useful tools in the homebrewer's toolbox.

When all of the individual modules are built, they should be mounted in a well screened enclosure. Make sure that each module is connected to a common ground. Connect your aerial to one side of the BPF. Connect the other side of the BPF to the RF input of the PD. Connect the VFO output to the local-oscillator port of the PD. The audio output from the PD module goes to the low level amp and finally to the audio power amp. The VFO and audio amplifier require a DC power supply between 9 and 14V. I have tested the receiver on my bench power supply at 13.8V and also with a 9V battery. When the receiver is up and running, try to find a strong station near the middle of the band and adjust L1, L2 and L3 for strongest received signal.

PERFORMANCE. If you have never heard a direct conversion receiver before, you will be quite surprised by the excellent audio quality produced by this little receiver. I have never heard such clear reproduction from a superhet receiver. Despite the lack of an RF amplifier, sensitivity is good enough to copy many DX and European stations. I use a half-wave dipole at about 11m above ground on 80m. This type of receiver gives no opposite sideband rejection. Any signals or noise on the opposite sideband will be heard along with the wanted signal. This receiver has not got an S-meter or any form of automatic gain control (AGC). The only way of coping with very strong signals is to manually reduce the AF gain. The lack of AGC is not a serious limitation, but it does require some skill on the part of the operator.

It is possible to build a DC receiver with single sideband selectivity. It is also possible to have proper AGC. However, the extra circuitry required would make the receiver a lot more complicated.

Audio samples of received signals can be found at:
http://homepage.eircom.net/~ei9gq/local.mp3
http://homepage.eircom.net/~ei9gq/dx.mp3

BPF MODULE COMPONENT LIST

C1, C7	680pF
C2, C6	180pF
C3, C5	22pF
C4	150pF
L1, L2, L3	nominal 10μH. 26 turns on Toko 10K type or similar slug tuned former. See text.

PD MODULE COMPONENT LIST

T1	5 turns, trifilar wound on ferrite toroid. Type 43 ferrite or similar ferrite core. Tip: the ferrite cup from inside Toko and similar inductors makes a good core for this type of transformer.
C8, C9	0.01μF
C10, C11	1μF
R1	47W
D1 to D4	4 x 1N4148 or similar high speed switching diodes.
L4	22μH
L5	100μH
L6	2.7mH

LOW LEVEL AUDIO AMP COMPONENT LIST

C27, C28, C30	10μF electrolytic
C29	1000μF electrolytic
C31	1μF electrolytic
C32	0.047μF
R10	47kΩ
R11	10kΩ
R12	4.7kΩ
R13	1kΩ
R14	3.3kΩ
R15	1.5kΩ
R16	120Ω
Q4	2SC945 or similar NPN general purpose transistor.
Q5	2SA733 or similar PNP general purpose transistor.

VFO MODULE COMPONENT LIST

C12	50pF variable capacitor. Preferably air-spaced and with a reduction drive
C13	150pF polystyrene or silvered mica. See text
C14, C15	1500pF polystyrene or other high stability type
C16	10pF
C17, C18, C25	0.1μF
C19	1000pF
C20	560pF
C21	100pF trimmer capacitor. A smaller value will do if you adjust the value of C20 accordingly.
C22	1500pF
C23	1000pF preferably a feed through type fitted where the +DC line enters the VFO enclosure.
C24, C26	0.01μF
R2	100kΩ
R3	100Ω
R4	82kΩnot critical. 100kΩwill work fine.
R5	560Ω
R6	680Ω
R7	5.6kΩ
R8	10Ω
R9	2.2kΩ
L7	11.5μH. 48 turns on a T50-2 toroid core. See text.
L8	100μH
L9	47μH
L10	3.8μH 28 turns on a T50-2 core
IC1	7805 5 volt regulator
D5	1N4148
Q1, Q2	BF245C FET
Q3	PN2222 or similar NPN transistor. Plastic package version of the 2N2222

Single Signal SSB/CW Receiver

I suffered from a lot of interference when I was listening in the DX window at the top of the 80m band using the simple direct conversion receiver we built. When listening to stations on LSB, the lack of opposite sideband suppression meant that I could hear any signals that were a few kHz above the wanted signal. Our old friend Murphy's Law says that any time you want to listen to a very weak DX or QRP signal, there will always be a very strong signal on the opposite sideband (image) frequency. The time has come to build a single signal SSB/CW receiver that will receive the wanted signal and reject the unwanted image signal. We will be able to re-use some of the blocks from previous projects for this new project.

The prototype receiver was built using dead-bug style construction. Dead-bug or similar methods of construction will make future modifications a lot easier.

The new receiver will be a superhet with a crystal IF filter. The superhet (Super Heterodyne) receiver converts the RF signal to an intermediate frequency (IF) where it is filtered and amplified before being converted to audio. A product detector is used for the IF-to audio conversion.

Crystal filters with suitable bandwidth for SSB and CW are prohibitively expensive. I have seen such filters advertised on eBay with an asking price of more than $100. That's more than my component budget for the entire year! I have some crystal filters in my junk box. A few minutes of digging produced a few 10.695MHz filters with a bandwidth of 2.7kHz. These filters were pulled from old CB rigs. I also found a high quality 9MHz filter with a bandwidth of 2.4kHz. If you have such a filter, you could use it in this receiver. I will resist the temptation to use one of my junk box filters because readers might be unable to find a similar filter. Instead, I will use a homebrew filter made from surplus crystals.

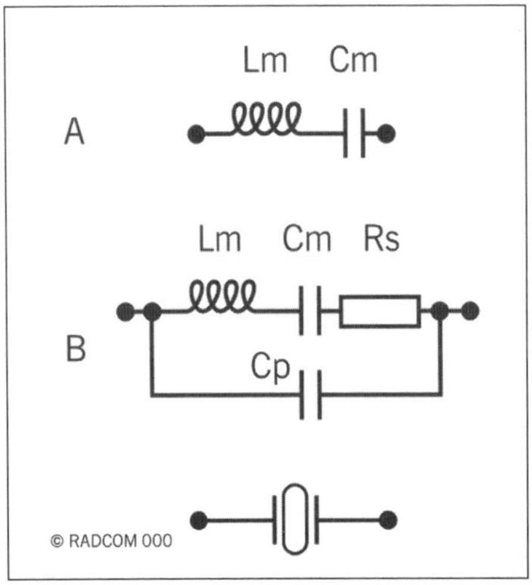

FIGURE 2.10
A: A simplified circuit of a crystal.
B: A more accurate model of a crystal.

THE QUARTZ CRYSTAL. The quartz crystal is a high Q resonant circuit. Its resonant frequency is determined by mechanical resonance rather than the reactances of an LC circuit. However, a quartz crystal behaves very much like a series resonant LC circuit. **Figure 2.10A** shows a simplified circuit of a crystal. Lm and Cm (motional inductance and capacitance) represent the values of L and C that would be found in an equivalent LC series resonant circuit. **Figure 2.10B** is a more accurate model of a crystal. Rs is a series resistance which represents losses in the crystal. Cp is the parallel capacitance of the

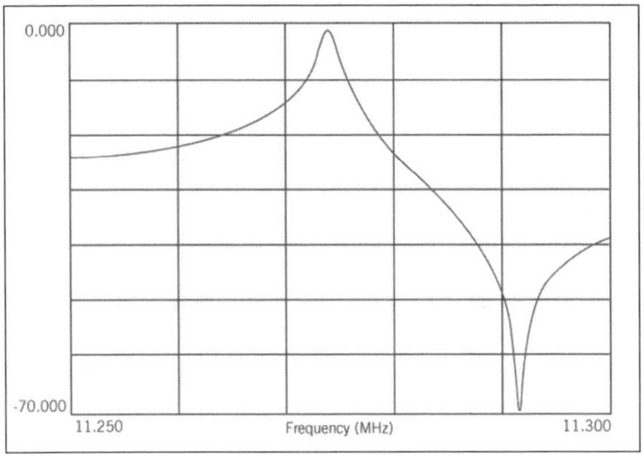

FIGURE 2.11: Frequency response of a typical crystal in a 50Ω test circuit.

crystal. Modern crystals have low Cp values of no more than a few pF. Cp can often be ignored when designing a ladder filter. The value of Cm is much smaller than you would find in conventional LC circuits. Cm is usually measured in femtofarads (1fF = 0.001pF). The value of Lm is quite high, usually more than 10mH. **Figure 2.11** shows the frequency response of a typical crystal in a 50Ω test circuit. The series resonant frequency is 11.274MHz. The very deep null at 11.292MHz is caused by the crystal's parallel resonance. This parallel resonant frequency, sometimes known as the anti-resonant frequency, is a result of the crystal's parallel capacitance (Cp).

THE CRYSTAL LADDER FILTER. There are good reasons why the crystal ladder filter is so popular with amateur homebrewers. All crystals in a ladder filter are resonant at the same frequency. You just need a handful of identical crystals and a few capacitors to make a high performance IF filter. The other common type of filter is called the crystal lattice. This type of filter is more difficult to design and build because it requires crystals that are resonant at two slightly different frequencies.

I will take the easy option and build a ladder filter. Crystals with a resonant frequency in the 6 to 12MHz range are often a good choice for an SSB bandwidth ladder filter. The traditional SSB IF frequency of 9MHz would be an ideal choice. Unfortunately, 9MHz crystals are not readily available on the surplus market. Cheap CPU clock crystals are available off the shelf for standard clock frequencies like 6, 8, 8.192, 10, 10.24, 11.192 and 12MHz.

When choosing an IF frequency, we should avoid frequencies that are likely to cause spurious responses in the receiver. Multiples of the VFO frequency should not be used. We should also avoid using an IF frequency that is too close to the amateur band/s we will be receiving. If the filter is to be used in a transmitter or transceiver, the choice of IF frequency is even more critical.

Before you start soldering, you should write down the RF, IF and local oscillator frequencies and spend a few minutes with a pocket calculator, looking for any likely causes of spurious signals. Apart from the more obvious problems like VFO harmonics, you should look out for third, fifth and perhaps even higher order intermodulation products, f1 ± 2xf2, f2 ± 2xf1, 2xf1 ± 3xf2 and so on.

CONSTRUCTION. **Figure 2.12A** shows a simple two-pole ladder filter. R is the filter input/output impedance. C is a shunt capacitor which is connected between each crystal and ground. **Figure 2.12B** shows how this scheme can be expanded to include any number of crystals. Unlike the lattice filter, a ladder filter can have an uneven number of crystals.

Now that we have chosen a suitable intermediate frequency and acquired a fistful of identical crystals, we can proceed to the easy bit. There are several ways of designing a ladder filter. We can design a filter empirically by making the filter first and worrying about the component values later. We could start with some reasonable value of C and R, say 120pF/150Ω or so. Once the filter is constructed, we can measure the frequency response of the filter and adjust the values of C and R until the desired response is obtained. This might seem like a haphazard way of designing a filter, but many excellent filters have been made this way.

A more scientific approach was described in 'A Unified Approach to the Design of Crystal Ladder Filters' by Wes Hayward, W7ZOI **[12]**.

Before we can calculate the correct values for the shunt capacitors and I/O impedance, we must accurately measure the crystal parameters. W7ZOI's article has detailed schematics of the homebrew test equipment he used to measure his crystals. If you have the necessary test equipment, you can make similar measurements using a signal generator, a few resistors and an oscilloscope or RF voltmeter. See **Figure 2.13** for a suggested test setup. If you don't have suitable test equipment, don't worry about it, you can still make a good filter by using empirical methods.

To test the crystals, the generator should be tuned slowly across the crystal resonant frequency. There will be a sharp peak in the measured voltage when the generator is tuned exactly to the series resonant frequency. This frequency should be measured carefully using a frequency counter, preferably to an accuracy of 10Hz or less. Next the generator should be tuned slowly downwards in frequency until the meas-

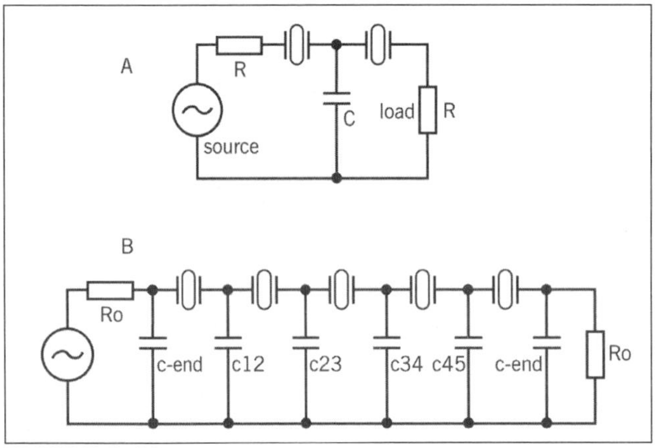

FIGURE 2.12
A: A simple two-pole ladder filter.
B: a longer ladder filter.

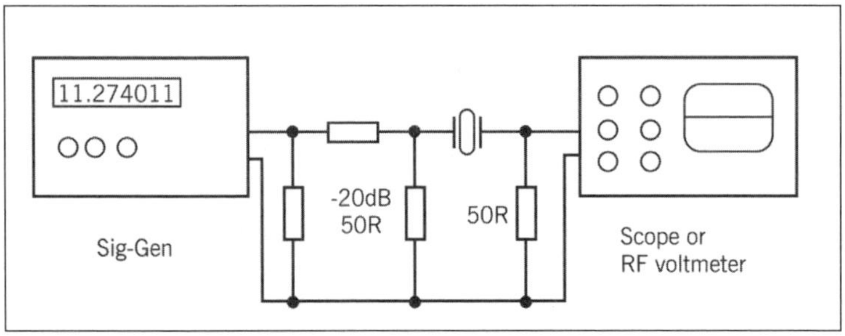

FIGURE 2.13: Set-up for testing the crystals.

ured voltage falls to 0.707 times the peak value which was measured at the series resonant frequency. This frequency should also be noted carefully.

Next the generator should be tuned upwards in frequency, back through the peak until the voltage again reaches 0.707 times the peak value. The difference between the highest and lowest frequency in Hz is the 3dB loaded bandwidth of the crystal when it is in the 50Ω test circuit. Remove the crystal from the test circuit and replace it with a variable resistor. A 47Ω potentiometer (pot) is ideal for this job.

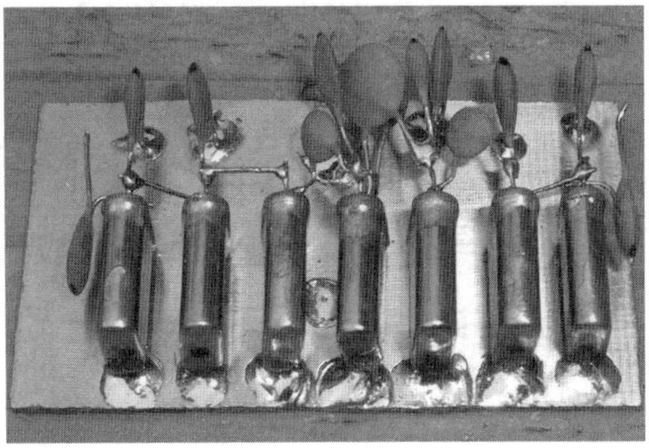

PHOTO 2.7: A ladder filter built using dead bug construction.

Tune the signal generator to the crystal resonant frequency and adjust the pot until the measured voltage output is exactly the same as the peak voltage measured when the crystal was in circuit. Remove the pot from the test circuit and measure its resistance. This will be the same as the crystal series resistance (Rs). Spend a little time making these measurements. Do the tests several times to make sure you are getting consistent results. You should test several crystals to make sure that they all have similar characteristics.

I have managed to acquire a fistful of 11.275MHz crystals for exactly the right price. I used one of these crystals to make a simple variable frequency crystal oscillator (VXCO) to use as a signal generator for the tests. A sensitive RF voltmeter (Marconi video voltmeter) was used for measuring the output from the test crystal. I found that the series resonant frequency of the crystals is 11.274MHz and that all crystals have the same resonant frequency to within ±100Hz. The measured 3dB bandwidth is 1,500Hz and the series resistance is 15Ω.

The following calculations will show the Qu, Lm and Cm values of the crystals.

Qu = (1.2 * 10^8 * F) / (f * Rs)
where F = series resonant frequency in MHz.
f = 3dB BW in Hz and
Rs is the measured series resistance.

In my case:

Qu=(1.2*10^8*11.274)/(1500*15)
=60,128
Cm = (1.326*10^{-15})*(f/F^2) Farads.
For my crystals: Cm=(1.326*10^{-15}) * (1500/11.274^2) = 1.564872E^{-14} Farads or if you prefer 0.01564872pF or 15.64872fF
Lm = 19.1/f Henries.
Lm = 19.1/1500 = 0.012733333 Henries = 12.733333μH.

Haywards formulas can now be used to calculate values for the coupling capacitors, the end capacitors and I/O impedance of a ladder filter. The value of each shunt capacitor in pF can be calculated from:

Cjk = 1326*(f/(B*Kjk*F))-10 pF where
f is the measured 3dB BW in Hz,
B is the desired filter bandwidth in Hz,
Kjk is a normalised coupling constant (see **Table 1**) and F is the crystal frequency in MHz

Let's see if we can design a five-pole filter based on the circuit in **Figure 2.12b**. The filter BW is 2500Hz. The filter will be terminated with 250Ω at both ends.

C12 and C45 = 1326*(1500/(2500*0.7028*11.274))- 10 = 90.4pF
C23 and C34 = 1326*(1500/(2500*0.5355*11.274))- 10 = 121.7pF

The end terminating resistance is calculated from:
Rend = (120*B) / (q*f) – Rs
(120*2500)/(1.3013*1500) – 15 = 138.7Ω
q is a normalised end-section Q from Table 1

Cend allows the filter to be terminated in a resistance that is higher than Rend
Cend = ((1.59*10^5)/(Ro*F)) * (√((Ro/Rend) - 1)) - 5
where Ro is the desired termination impedance. We want to terminate our filter with 250Ω so Ro = 250
Cend = ((1.59*10^5)/(250*11.274)) * (√((250/138.7) -1)) -5 = 45.54pF

I was never any good at maths. I managed to do the above calculations on my Sharp scientific calculator. I have made a spreadsheet to do the calculations automatically. The spreadsheet can be found at:

http://homepage.eircom.net/~ei9gq/filter.html

Figure 2.14 shows a plot of the calculated frequency response of the five-pole filter. Plots for a six and seven-pole version are shown for comparison. The plots were made with ViPEC network analysis software **[13]**. The ViPEC input files are also available for anybody who wants them. The measured response of the actual filter is remarkably close to the response predicted by ViPEC. The passband ripple of this filter is a little more than the specified value of 0.1dB.

This is largely due to the detuning effect of the shunt capacitors. For example, the crystal in the middle of the filter has 127pF of shunt capacitance at each end. The crystals at each end have got 90pF at one side and just 45.5pF at the other.

This unequal detuning of the filter crystals means that some crystals will resonate at a slightly different frequency to the others. The first and fifth crystals will resonate at the same frequency, the second and fourth crystals will both resonate at a slightly different frequency and so on... This small error can easily be corrected by using a capacitor in series with some or all of the crystals. As the error only causes a passband ripple of around ±1dB, we can safely ignore it for now. You can use standard value capacitors connected in parallel to make up the values required. I used 39+6.8=45.8, 82+8.2=90.2 and 100+27=127. This is near enough to the required values

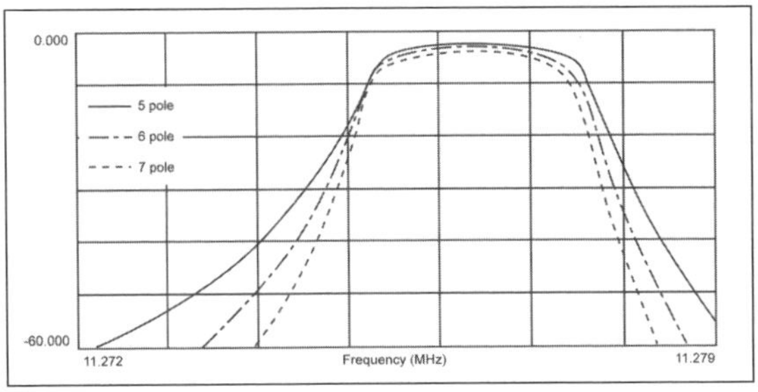

FIGURE 2.14: Plot of the calculated frequency response of the five-pole filter.

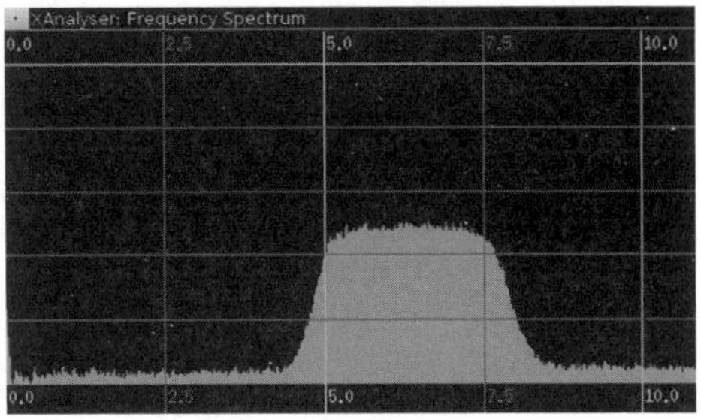

FIGURE 2.15: A spectrograph of the noise measured at the output of a homemade seven-pole ladder filter.

It is possible to measure the passband shape of a crystal filter by using a noise source and a spectrum analyser. You might have a spectrum analyser in the shack without being aware of it. A computer with sound input can be used to sample audio and analyse its spectrum. The product detector and low level audio amplifier from a previous project can be used to convert the noise from the filter output to audio frequencies so that it can be sampled.

This combination of a simple RF to AF converter and a PC makes a very useful spectrum analyser. It can only look at a 22kHz wide slice of spectrum at any one time, but this is more than adequate for this task.

You only need a crystal oscillator that is running at a few kHz above or below the filter passband. Perhaps you have already built one for the tests we did earlier. This oscillator is used as the local oscillator for the mixer/product detector. The noise from the noise source is fed through the filter under test to the RF port of the product detector. The resulting audio output from the product detector is amplified by the low level audio amplifier before it is sent to the line-in input of the PC sound card.

For best results, the audio low pass filter following the product detector should be disabled by removing C11 and replacing L6 with a wire link. **Figure 2.15** shows a spectrograph of the noise measured at the output of a homemade seven-pole ladder filter. This measurement shows that the filter has a -6dB bandwidth of less than 2.5kHz. The frequency markers are 5kHz per division. The amplitude scale is 20dB per division. The bandwidth at about -40dB would suggest the predicted 60/6dB shape factor of better than 2:1 has been achieved. A wonderful program called Xanalyser **[14]** was used for the spectrum analysis **[15]**.

N	q	k_12	k_23	k_34	k_45	k_56	k_67	k_78
3.00	1.4328	0.6618	0.6618					
4.00	1.3451	0.6850	0.5421	0.6850				
5.00	1.3013	0.7028	0.5355	0.5355	0.7028			
6.00	1.2767	0.7145	0.5385	0.5180	0.5385	0.7145		
7.00	1.2615	0.7223	0.5421	0.5155	0.5155	0.5421	0.7223	
8.00	1.2515	0.7276	0.5451	0.5160	0.5100	0.5160	0.5451	0.7276

TABLE 1: Normalised coupling constants for an n-pole 0.1dB ripple Chebyshev filter.

A Simple Spectrum Analyser

Some of the previous modules can be used as the frontend of a spectrum analyser (**Figure 2.16**). The mixer/product-detector was changed from a single balanced type to a double balanced mixer (DBM). This change only requires the addition of a single component: T2. The construction of T2 is identical to T1 from the original design. The audio low pass filter from the DC RX has a cut-off frequency of about 3kHz. This would severely limit the bandwidth of the spectrum analyser. This problem can be overcome by disabling the low pass filter (LPF). The easiest way of doing this is to remove C11 from the circuit and replace L6 with a wire link. Alternatively, you can raise the LPF cut-off frequency to 24kHz by changing the value of C10 and C11 to 0.133µF and L6 to 330µH. As 0.133 is not a standard capacitor value, you can use 0.1 in parallel with 0.033µF. The new component values are shown in **Figure 2.17**. I used ordinary 1N4148 switching diodes in the mixer. The AF amplifier block in Figure 2.16 is the low-level amplifier from the DC receiver project. The high frequency response of this amplifier can be improved by changing C32 from 0.047µF to 0.01µF.

Before we can use the modified frontend to measure the response of our crystal ladder filter, we need to build a new local oscillator and a broadband noise generator. **Figure 2.18** shows the schematic of the noise generator. A Zener diode is used as the noise source.

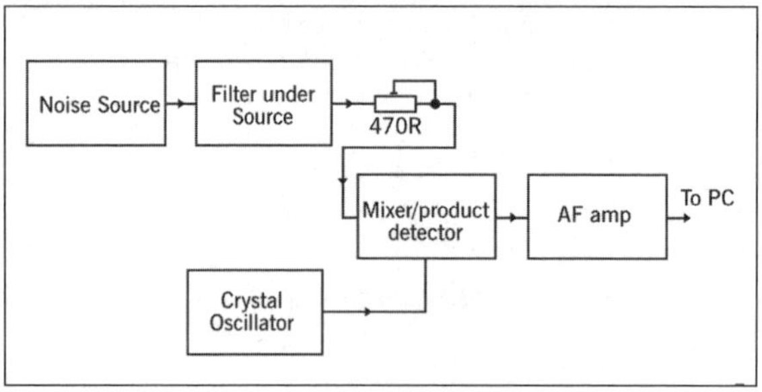

FIGURE 2.16: Block diagram of the spectrum analyser.

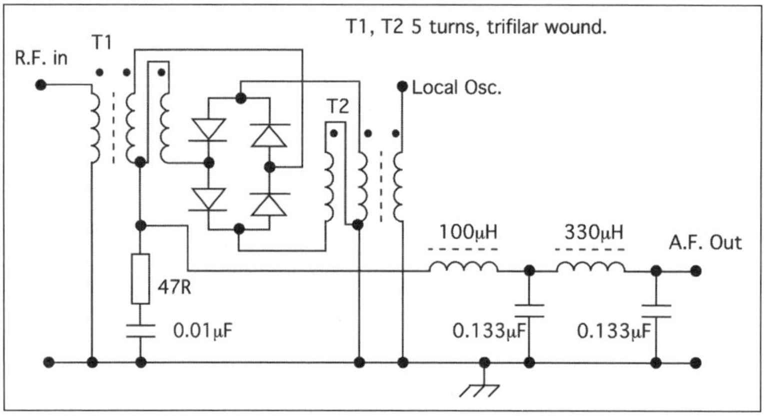

FIGURE 2.17: The modified product detector.

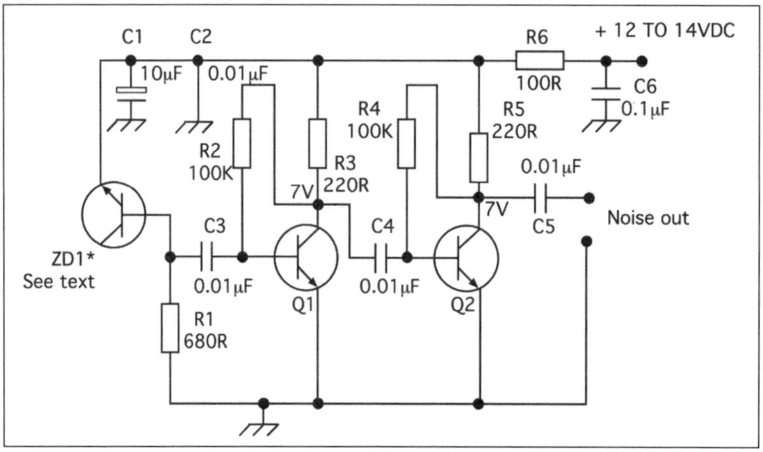

FIGURE 2.18: Schematic of the noise generator.

When a Zener diode is reverse biased to the point where it reaches its reverse breakdown voltage, it produces a relatively high level of noise. This noise is amplified to a level of several hundred mV by a pair of transistor amplifiers. I used the base-emitter junction of a transistor in place of the Zener diode. **Photo 2.8** shows the layout of my circuit.

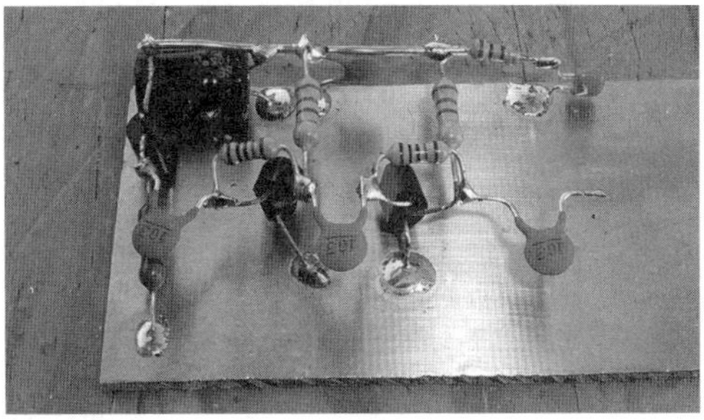

PHOTO 2.8: Dead bug construction of the noise generator.

The schematic of the new local oscillator is shown in **Figure 2.19**. Q3 is configured as a common collector Colpitts oscillator. Q4 is an emitter follower buffer amp. The oscillator should run a few kHz above or below the centre of the IF filter passband. I used one of my surplus 11.275MHz crystals in this circuit, see **Photo 2.9**. My ladder filter has a centre frequency of 11.276MHz. I need to convert the RF noise at the output of the ladder filter down to audio frequency so that it can be sampled by my personal computer. To achieve this, I need to have the oscillator running at 11.276MHz plus or minus a few kHz.

PHOTO 2.9: Construction of new local oscillator.

It is very unlikely that you will have a spare crystal that is resonant at a frequency that just happens to be about 5kHz above or below your filter passband. For instance, if you use 10MHz CPU clock crystals for your ladder filter, you might expect to find the centre of the filter passband at 10.001MHz. This would require a 9.996 or 10.006MHz crystal for the local oscillator.

The frequency of a crystal oscillator can be 'pulled' by placing a capacitor or inductor in series with the crystal. A small amount of series capacitance would effectively be in series with the crystal's motional capacitance (Cm). This would reduce the value of Cm and increase the resonant frequency of the crystal. Similarly, adding a small amount of inductance in series with the crystal will increase the effective value of motional inductance and lower the frequency of the crystal.

If we ensure that the reactance of the external L or C is very small when compared to the

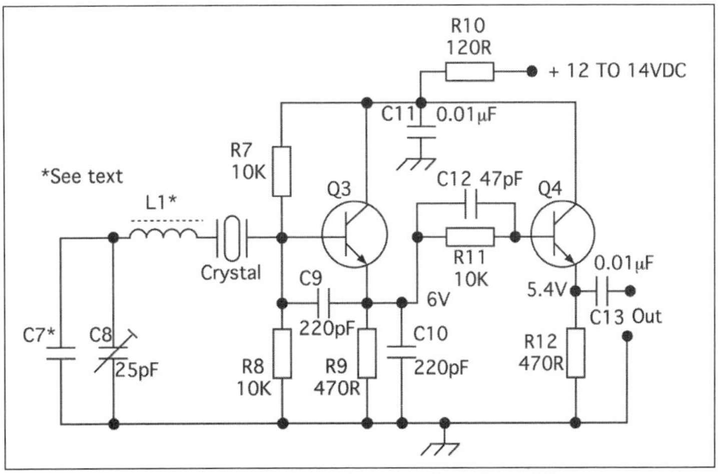

FIGURE 2.19: Schematic of the new local oscillator.

reactance of the crystal's Cm and Lm, the Q and stability of the crystal will only be degraded by a small amount. As a very rough rule of thumb, a typical crystal can be pulled by a maximum of 0.1%, or about 10kHz for a 10MHz crystal. I tried testing a few 8, 10 and 12MHz crystals from the junk box to see how far I could pull them off frequency. All of them ran quite happily at more than ±5kHz.

Note that C7 and L1 in Figure 2.19 are optional components. If you do need to use a series inductance to pull the oscillator to a lower frequency, a typical value for L1 would be somewhere between 2 and 20µH. I used a 6.8µH choke with my 11.275MHz crystal.

A typical value for C7 is around 22pF. To find the highest frequency at which a crystal will oscillate, remove L1 and C7 and adjust C8. Spend a little time experimenting with the component values in the crystal oscillator. Time spent learning how to tweak the frequency of a crystal oscillator now might save you the price of a pair of crystals when we get to the carrier oscillator (CIO) stage of this receiver.

A good frequency counter or a receiver with a frequency display is a useful tool for this kind of work. If you don't have a frequency counter, don't rush out and buy one just yet, we will be building a counter in the not too distant future.

Now that all of the modules have been built, we can assemble the test rig for our ladder filter. The output from the noise generator should be connected to one side of the crystal filter. The other side of the crystal filter connects to the RF port of the product detector. Note that the product detector has an input impedance of around 50Ω. A series resistor of about 200Ω is required, so that the filter output will be properly terminated. I used a 470Ω pot for this purpose. The output from the new local oscillator is connected to the LO port of the product detector. The audio frequency output from the product detector module goes to the low-level audio amplifier and finally to the line-in socket of the PC sound card. Use screened cable for the connection to the PC. Each module should be connected to a common ground and well screened from any source of RF noise.

All we need now is a filter to test. I used the spreadsheet that I mentioned previously (http://homepage.eircom.net/~ei9gq/filter.html) to design an eight-pole SSB filter. The filter bandwidth is 2,500Hz. I used the nearest standard values for the shunt capacitors. For the 130pF capacitors, I used 120pF in parallel with 10pF. **Figure 2.20** shows the schematic of the filter. I used eight carefully matched 11.275MHz crystals.

When I started up the spectrum analysis software, I was hoping to see a nice plot of the filter's frequency response. Instead, I found myself looking at an empty spectrum display! This problem was solved

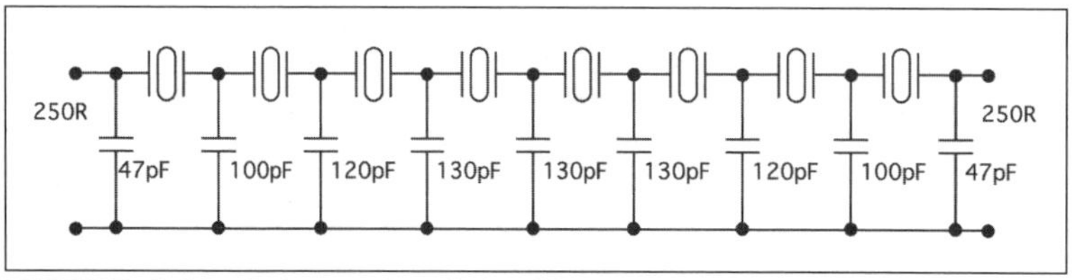

FIGURE 2.20: Schematic of the filter.

by changing the sound card input device from 'Mic' to 'Line' and adjusting the input gain control to an appropriate level. Make sure that you are adjusting the control for the input gain and not the output gain.

Once the sound input was configured correctly, I was able to see the shape of the filter passband on the spectrum display.

During the first test, the top of the filter passband was only 40-45dB above the noise floor. Most of the noise was coming from my PC monitor. Putting the entire circuit in a screened enclosure (I actually used a biscuit tin, but 'screened enclosure' sounds so much better) reduced the noise floor to 60dB down the slope of the filter's response. I was quite pleased to find that the response of the filter was very close to ideal for an SSB filter. There was no need to make any changes to the component values.

So, at last, the moment of truth. How good is my ladder filter? To find out, I tested several SSB filters from the junk box to see how they compared to my homebrew unit. The six-pole filters I recovered from dead CB rigs were surprisingly good. Several of these filters had a bandwidth of about 2.5kHz and a shape factor of about 1:2.2, not bad for a six-pole filter. One ex-CB eight-pole filter had a bandwidth of 2.4kHz and a shape factor that was just as good as the homebrew filter. Finally, I tested a Yaesu Musen XF-8.2M-242-02 filter. This is a 2.4kHz BW, eight-pole SSB filter. **Figure 2.21** shows the measured response of the homebrew filter and the Yaesu unit.

The ladder filter has a measured bandwidth of 2.6kHz, which is slightly more than expected. The bandwidth of the Yaesu filter is very close to the manufacturer's specification of 2.4kHz. Both filters have a 6/60dB shape factor of about 1:1.85. The spectrum analysis plots were made using Xanalyser running on my Linux PC. Note that Xanalyser's Y-axis (vertical) grid is 20dB per division. There are similar spectrum analysis programs available for Windows and MacOS.

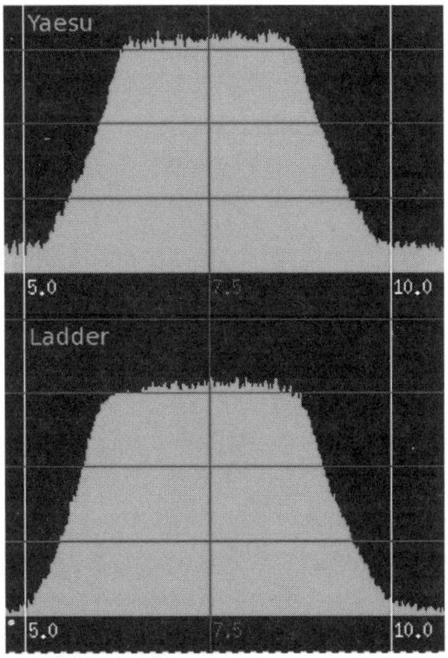

FIGURE 2.21: The measured response of the homebrew filter (bottom) compared with a 2.4kHz BW eight-pole SSB filter (top).

I spent a little time with a signal generator, an RF millivolt meter and an oscilloscope to see if I could measure the ultimate attenuation of the filter for signals that are outside the passband. At ±2.6kHz from the centre of the filter passband, the signal was already 90dB down. There was no evidence of any spurious responses above this level when the generator was tuned well above and below the filter passband. This compares very well with even the highest quality commercially made filters.

This type of filter should attenuate out of band signals by well over 100dB. I was unable to measure this level of attenuation with my simple test equipment. Even if such extreme rejection of out of band signals is theoretically possible, it would be difficult if not impossible to build a circuit that would provide such a high level of isolation between the filter input and output.

There will always be some small amount of signal leakage around the filter. If the crystal oscillator is replaced by a variable frequency oscillator (VFO), this simple tool can be used to look at any 22kHz wide slice of spectrum from LF to VHF **[16]**.

IF AMPLIFIER. The next stage in our homebrew receiver is the IF amplifier. Most of the receivers I have built in the past have used one or more ICs in the IF amplifier stages. The most popular devices like the SL1612 and the MC1350P are no longer in production. I have a choice of using a more modern device like the AD603 or using discrete transistors for the IF amplifier. I want to use components that are readily available so I have decided to use dual-gate MOSFETs in the IF amplifier stages.

The classic dualgate MOSFET IF amplifier has the IF signal applied to gate 1 and the AGC (automatic gain control) voltage applied to gate 2 of the device. MOSFETs that are designed for use as RF amplifiers in VHF/UHF TV tuners make excellent IF amplifiers. This type of amplifier will have about 25-30dB of gain and an AGC range of approximately 40dB. Low noise devices like the BF966S are readily available in surface-mount packages. The BF981and BF961 are available in TO-50 (pill type) packages. These devices cost well under £1 each.

To get the maximum gain control range, these devices must be biased so that gate 2 of the device has a negative voltage relative to the source pin. Maximum gain will be achieved with 3 to 4V DC on gate 2. Minimum gain (maximum attenuation) requires minus 2V DC on gate 2. The prototype IF amplifier was built dead-bug style on copper-clad PCB.

Figure 2.22 shows the schematic of one of the IF stages. The schematic only shows a single IF amplifier and product detector. A pair of identical IF amps will be used in the finished receiver. The product detector is identical to the one used in the direct conversion RX project.

PHOTO 2.10: My partially built amplifier.

A pair of high frequency amplifiers with a combined gain of 50 to 60dB, a tuned circuit in each amplifier stage and a high-Q crystal filter connected to the amplifier input provide a perfect recipe for instability. Great care should be taken with the construction and physical layout of the circuit. Component leads should be no longer than necessary. The circuit should be laid out so that there is good isolation between input and output. The pin-out of the BF981 and similar devices is designed to help you achieve this goal. I have built two of these IF strips and so far, I haven't had any problems with instability. The photograph (**Photo 2.10**) of my partially built amplifier shows the layout I used. Note the short leads on the decoupling capacitors. The source terminal of the MOSFET is held at about 2V above ground by the voltage drop across the LED. This gives a much greater range of gain control when AGC is applied to gate 2 of the MOSFET.

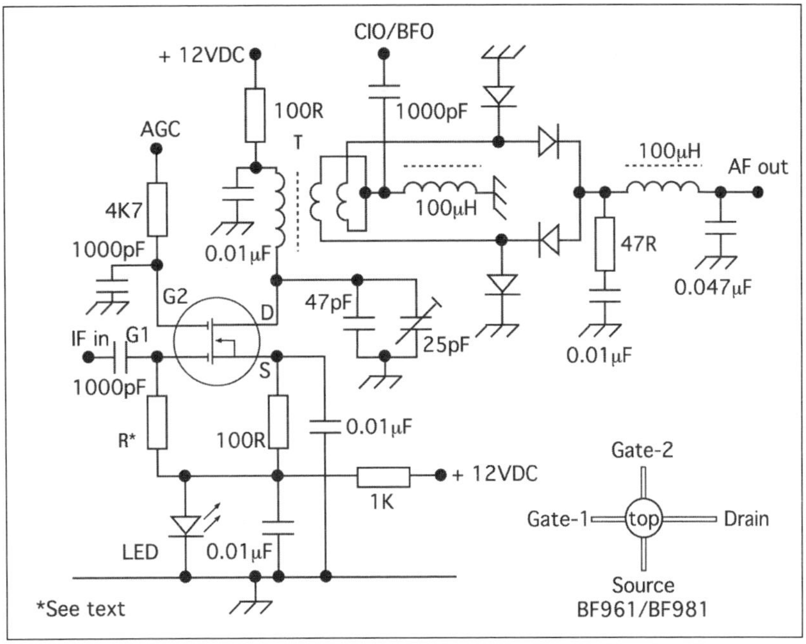

FIGURE 2.22: The schematic of one of the IF stages for the IF amplifier project.

The value of the resistor R* is chosen to provide the desired impedance at the amplifier input. My crystal filter needs a 250Ω termination. My first prototype of the IF amplifier had a 470Ω pot in this position. When I was testing the amplifier, I found that the best passband shape was achieved when the pot was adjusted to about 250Ω. I always find it a little surprising when things just work out exactly as they are supposed to! The pot was later replaced by the nearest standard value resistor of 270Ω .

Instead of using a standard IF transformer in the drain circuit of the MOSFET amplifier, I decided to use a home made transformer. The core is T68-2 powered iron toroid. I used 27 turns of enamelled copper wire as the primary winding and four turns bifilar wound as the secondary winding. To make the bifilar winding, use two lengths of wire twisted together. **Figure 2.23** shows the arrangement of the windings. Wire A is the primary. Wire B and C are twisted together to form the secondary winding. One end of wire B is connected to the opposite end of wire C to make the centre-tap of the secondary winding. This produces the necessary antiphase outputs to drive the diodes in the product detector.

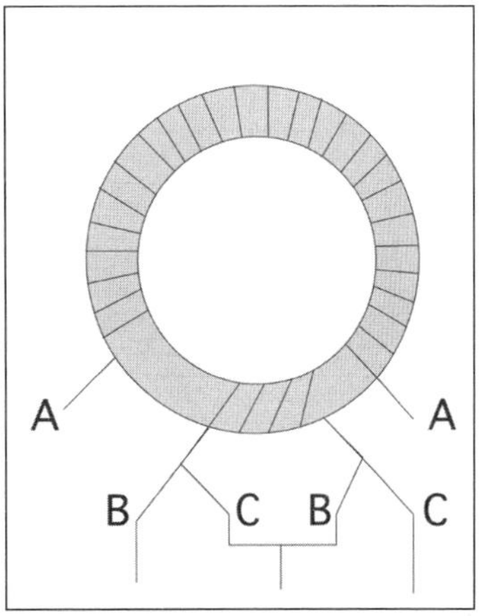

FIGURE 2.23: The arrangement of the windings for our homebrew IF transformer.

According to the Amidon data for the T68-2 [17], 27

turns will result in an inductance of 4.15µH. The standard formula for inductive reactance (XL = 2*PI*f*L) tells us that this inductor will have a reactance of 294Ω at 11.275MHz. It will require a parallel capacitor with the same capacitive reactance to make a parallel resonant circuit: C = 1/(2*PI*f*XC) gives us a value of 48pF.

I decided to use a 39pF fixed capacitor in parallel with a small trimmer with a maximum capacitance of 20 to 25pF. To tune the IF transformer to 11.275MHz, I found that I needed to adjust the trimmer capacitor to almost maximum capacitance.

This error is probably due to the fact that the 39pF capacitor is a cheap 10% tolerance ceramic disc type. Increasing the fixed capacitor to 47pF meant the trimmer capacitor was in the middle of its tuning range when the IF amplifier was tuned correctly.

These values are correct for my 11.275MHz IF frequency. The circuit will need little or no modification for a 10.7MHz IF. The circuit can easily be modified to work at any IF frequency from 8 to 12MHz by changing the value of the fixed capacitor.

NOISE GENERATOR COMPONENT LIST

C1	10µF electrolytic
C2, C3, C4, C5	0.01µF
C6	0.1µF
R1	680Ω
R2, R4	100k
R3, R5	220Ω
R6	100Ω
ZD1	base-emitter junction of 2SC945 or similar NPN transistor.
Q1, Q2	2SC945 or similar general purpose NPN.

CRYSTAL OSCILLATOR COMPONENT LIST

C7	22pF see text
C8	trimmer capacitor 25pF approx.
C9, C10	220pF
C11, C13	0.01µF
C12	47pF
R7, R8, R11	10k
R9, R12	470Ω
R10	120Ω
Crystal – see text.	
L1	2 to 20µH see text.
Q3, Q4	2SC945 or similar general purpose NPN.

IF AMPLIFIER COMPONENT LIST	
C1, C2, C10, C16	1000pF
C3, C4, C7, C9, C11, C12, C13, C17	0.01μF
C5, C14	47pF
C6, C15	25pF trimmer
C8	10pF
C18	0.047μF
R1	270Ω* to match filter Z
R2, R6	4k7
R3, R4, R5, R8	100Ω
R7	100k
R9	1k
R10	47Ω
D1-D4	1N4148 or similar
D5	LED
Q1, Q2	BF981 or similar dual gate MOSFET (BF961)
L1	27 turns on T68-2 toroid
T1	27 turns pri. 4 turns bifilar sec. wound on T68-2 toroid
L2, L3	100μH

Diode Ring Mixer

The schematic of a finished two-stage amplifier and product detector is shown in **Figure 2.24 (overleaf)** and **Photo 2.11**. The two stages are almost identical. The IF transformer (T1) of the second stage has a secondary winding to connect to the diodes in the product detector. The input resistor of the first stage (R1) is chosen to match the output impedance of the crystal IF filter. The amplifier has a gain of 50 to 60dB and AGC control range of more than 80dB (Photo 2.10).

Before we can do anything useful with our IF amplifier, we will need an RF frontend, a carrier oscillator (BFO/CIO) and audio amplifier stages. Most of the bits we need have already been designed and built for earlier homebrew projects. **Figure 2.25 (overleaf)** shows the block diagram of the complete receiver. Approximate gain/loss figures are shown for each stage in the signal path.

PHOTO 2.11: The diode ring mixer.

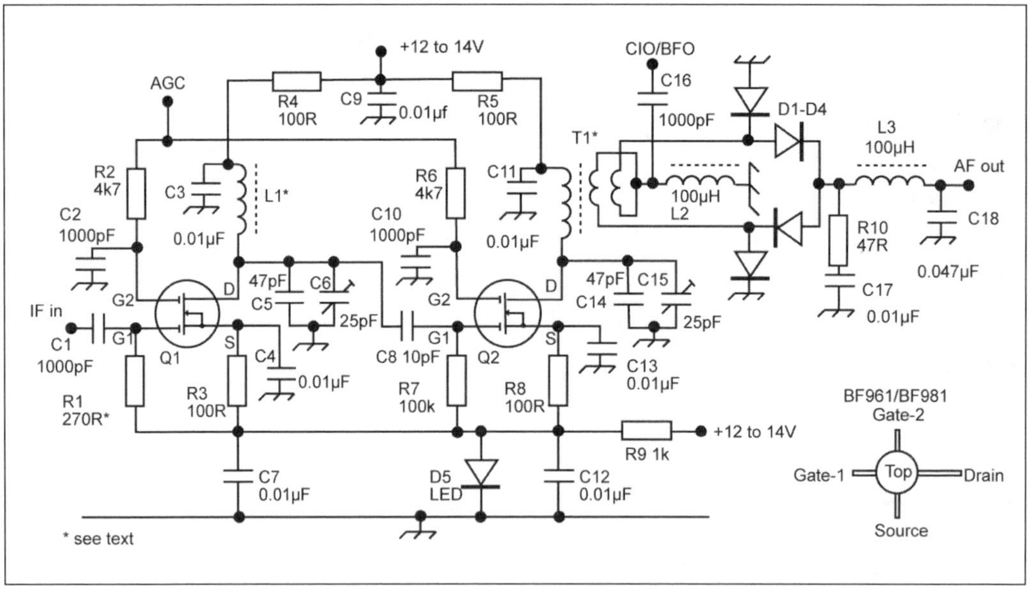

FIGURE 2.24: The schematic of the two-stage amplifier and product detector.

RECEIVE MIXER. The double-balanced diode mixer as used in the spectrum analyser project would also make a very good first mixer for our superhet receiver. This type of mixer is commonly known as a DBM (not to be confused with dBm) or diode ring mixer. The first mixer is a very critical part of a superhet receiver. The mixer is before the crystal IF filter in the receive signal path. This means that the mixer and any RF or post-mixer amplifiers in the receiver frontend are not protected from strong signals that are within the passband of the RF band pass filter. Intermodulation distortion (IMD) caused by the mixing of several strong signals may lead to unwanted distortion products appearing at the receiver's IF frequency. The 40m amateur band presents a worst case scenario where there are many high powered broadcast stations inside the BPF passband. If we are to avoid interference from unwanted IMD products, we must use a mixer with good strong signal handling capabilities.

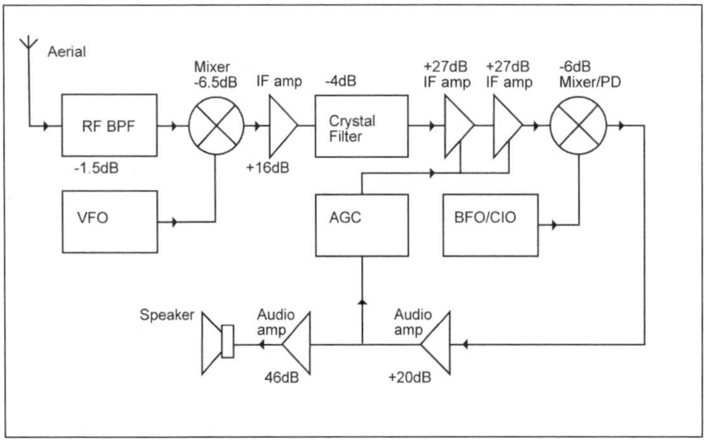

FIGURE 2.25: Block diagram of the complete receiver.

DYNAMIC RANGE. A receiver's two-tone dynamic range is the ratio between the weakest signal it can receive (MDS) and the strongest signals it can handle without generating unwanted distortion products. The MDS or minimum discernible signal is fairly easy to quantify. It is usually defined as a signal that is 3dB above the noise floor of the receiver. At the other end of the scale, signals that are strong enough to produce IMD products that rise 3dB above the noise floor are the strongest signals that can be received without generating unwanted spurious signals. The ratio between these two levels is knows as the spurious free dynamic range or SFDR.

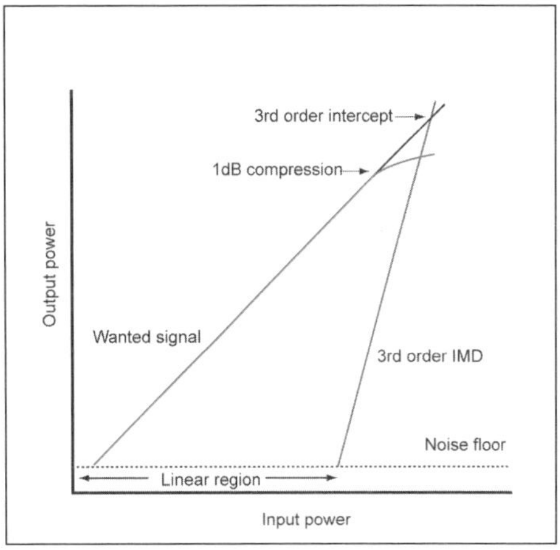

FIGURE 2.26: The relationship between input signals and third order intermodulation products seen at the output.

COMPRESSION POINT. A linear mixer or amplifier will have a limit to the maximum amount of output power it can deliver. As the output of the device is increased, it will eventually reach a point where its gain begins to decrease, or in the case of a passive diode mixer, its loss begins to increase. When this gain reduction amounts to 1dB, the device has reached its 1dB compression point. Driving a linear amplifier or mixer beyond this point will result in very high levels of distortion. The 1dB compression point should be regarded as an absolute upper limit. A mixer or amplifier in the front-end of a receiver will usually operate at a much lower level. The 1dB compression point is a useful indicator of the strong signal handling capabilities of a mixer or amplifier.

THIRD ORDER INTERCEPT. When a mixer or amplifier is operating in its linear region, the ratio of input power to output power (gain) remains constant. If the input level is increased by 10dB, the output level will also increase by 10dB. Unfortunately, IMD products have a tendency to increase at a much higher rate than the input signals. The graph in **Figure 2.26** illustrates the relationship between input signals and third order intermodulation products seen at the output. The third order products 2*f2-f1 and 2*f1-f2 are particularly troublesome because they tend to be close to the frequency of the wanted signal. Third order products increase at the cube of the input power, or to put it another way, for each 1dB increase in input signals, the third order IMD products will increase by 3dB. This means that putting an RF preamp with 20dB of gain before the mixer will increase wanted signals by 20dB and unwanted third order IMD by 60dB!

The two-tone third order intercept of a mixer or amplifier can be calculated by applying two signals to the input and checking the level of third order IMD products at the output. Two signal generators are needed to generate the test tones. A spectrum analyser is used to measure the levels of IMD. Since the third order IMD rises at a faster rate than the wanted signals, there must be a level where the IMD products will be just as strong as the wanted signals. This hypothetical point is known as the third order intercept point, sometimes referred to as IP3 or TOIP. This level is never reached in practice but knowledge of the intercept point can be used to calculate the amount of IMD we can expect to find at any level within the linear range of the device. If a mixer were to have an output intercept point of +20dBm (100mW) we know that if the same mixer is operated with an output of 0dbm (1mW) the third order products will be reduced by 60dB. The third order IMD products are now 40dB below the wanted signals. Reducing the input signals by a further 20dB will reduce third order products by another 60dB and so on until the distortion products disappear beneath the noise floor.

We must be careful when interpreting manufacturers' claims for dynamic range. There are many different ways of measuring DR, each method will result in a different figure. Some manufacturers specify the blocking dynamic range which is the ratio between the MDS and the level at which a single off-frequency signal will reduce the receiver gain by 1dB. This ratio will be much higher than the two-tone DR. The spacing between the two test signals can have a profound effect on two-tone DR test results. Some rig manufacturers have used test tone spacing of up to 50kHz for DR testing. Testing with tone spacing of 5 to 10kHz would be a more stringent test **[18]**.

I have chosen to use a diode DBM as the first mixer in our superhet receiver. There are a few other options available. I could have used a high-level mixer IC like the AD831 or a mixer built from discrete JFETs or MOSFETs. The modular construction of the receiver allows for easy changing of the first mixer. You can easily substitute another type of mixer to see how well it performs. The diode DBM offers excellent strong signal handling and can be built using readily available components. The diodes are BAT43 **[19]** Schottky signal diodes. These devices have a switching time of 5ns and a lower turn on voltage than standard silicon switching diodes.

Ordinary switching diodes like the IN914 or 1N4148 can be used instead. The transformer cores are type 43 ferrite toroids. Each transformer has six turns of trifilar wound insulated wire. I used three lengths of standard telephone wire twisted together. The coloured insulation makes it very easy to identify the wire ends. See **Figure 2.27** for details. If small type 43 or similar ferrite toroid cores are not available, you can use the ferrite 'cups' from Toko 10K or similar type inductors.

RULES OF THUMB. The input 1dB compression point of a diode DBM is usually a few DB below the LO injection level. Typically around 10 to 14dBm for a mixer with +17dBm of LO power. The conversion loss of a typical DBM is about 6 to 7dB. The third order intercept point of a diode DBM is typically 10 to 15dB above the 1dB compression point. The output 1dB compression point of a Class A amplifier usually coincides quite closely with the point where the peak signal current is equal to the DC standing current. To get the best performance from a DBM, the RF, LO and IF ports must be properly terminated. This usually means that all ports should be terminated with a 50Ω resistance. This is a lot more difficult than it sounds. Proper termination is required, not just at the RF, LO and IF frequencies, but also at the IF image frequency and any other frequency that may

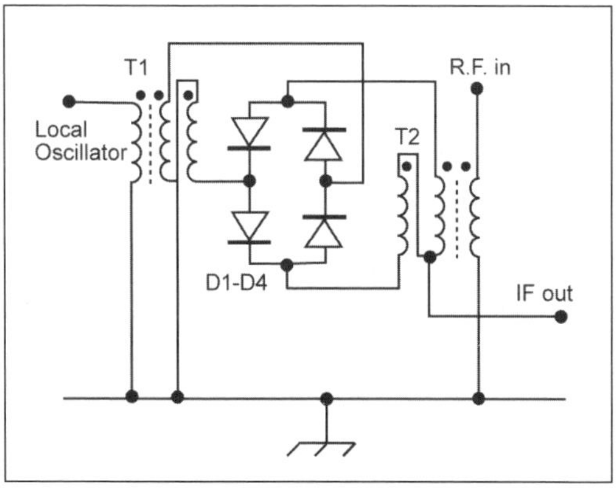

FIGURE 2.27: Each transformer in the diode ring mixer has six turns of trifilar wound insulated wire.

appear at the ports. This ideal situation is rarely achieved in practice.

Termination of the IF port is particularly critical. There are a few standard ways of terminating the IF port. Using a 50Ω resistive attenuator will improve the match at the IF port. This approach will increase mixer losses that will increase the system noise figure. Another approach is to use a broadband amplifier at the IF port. The amplifier must have a 50Ω input impedance over a wide bandwidth. The third and probably the best solution is to use a 50Ω diplexer at the mixer IF port.

The diplexer is a tuned LC filter that allows signals at the IF frequency to pass with very little attenuation. Signals at other frequencies are diverted into a 50Ω resistor. I opted for the broadband amplifier approach. I left some empty space between the mixer and post-mixer amp so that I have the option of adding a diplexer if I think it would improve the performance of the receiver.

POST MIXER AMPLIFIER. The first IF amplifier in this receiver is placed between the mixer and the crystal IF filter. This arrangement is commonly known as a post-mixer-amplifier. It might seem more logical to place this amplifier after the IF filter where it would be protected from strong signals on adjacent frequencies. However, there are at least two good reasons for placing it directly after the mixer. The combined loss of the RF band pass filter and the first mixer is about 8dB.

Adding a further 4 to 6dB of IF filter insertion loss before the first stage of amplification might increase the receiver's overall noise figure by an unacceptable amount. The resulting loss of sensitivity could only be overcome by using an RF preamp. In this case, the use of a post mixer amp is usually the lesser of two evils. The post mixer amp also serves as a buffer between the IF port of the mixer and the very reactive load presented by the crystal IF filter.

Isolation between the mixer IF port and IF filter is further improved by the use of a resistive attenuator between the amplifier and IF filter. The gain of the amplifier is around 20dB. About 4dB is lost in the attenuator, giving an overall gain of 16dB for the post mixer amp module. The attenuator is designed to match the IO impedance of the crystal filter. I used 4dB instead of the more common arrangement of 3 or 6dB so that standard value resistors could be used.

The amplifier (**Figure 2.28**) was built using dead-bug style construction on a scrap of cop-

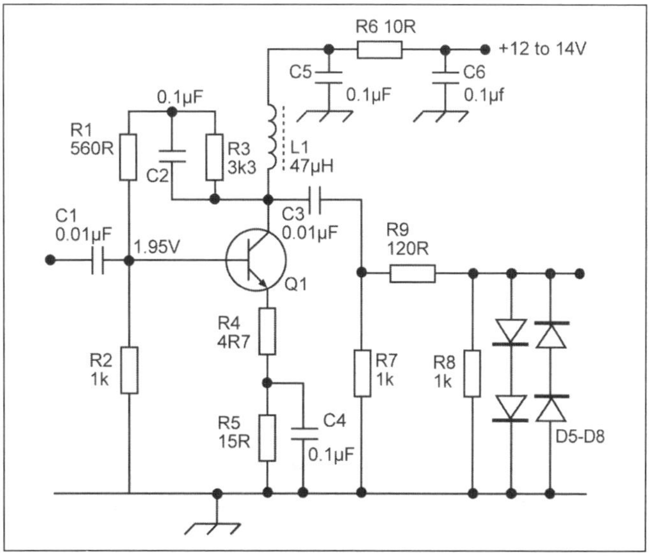

FIGURE 2.28: Schematic of the post mix amplifier.

PHOTO 2.12: The post mixer amplifier was built using dead-bug style construction on copper-clad PCB.

per-clad PCB. **Photo 2.12** shows the layout used. The 47μH inductor should be a high power type rather than a miniature or surface mount choke. If such an inductor is not available, you can use 11 turns on an FT37-43 or similar ferrite toroid.

Because the amplifier uses shunt feedback from collector to base (R1) and emitter degeneration (R4) to control gain and I/O impedance, it is unconditionally stable. The transistor takes about 63mA from the DC supply. With a supply voltage of 13.8V, the power dissipation in the transistor is about 750mW. A small push-on heatsink as seen in the photograph seems quite adequate as I sit here in my very chilly outdoor shack, listening to the hailstones rattling on the roof. I will probably invest in a larger heatsink before Summer. A VHF/UHF power transistor is used for Q1. I tested the amplifier with 2N3866, 2N3553 and BFW16A transistors. All three devices gave broadly similar results. The measured 1dB output compression point was better than +20dBm (100mW) for all versions. Modern surface-mount devices like the BFG235 would probably give even better results.

The post mixer amp looks more like a driver stage for a QRP transmitter than a low-level receive amplifier. However, even this rather sturdy amp will have overloaded before the diode mixer reaches its output 1dB compression point! The four diodes at the output of the post-mixer amp will protect the crystal filter from very strong signals at the amplifier output. I

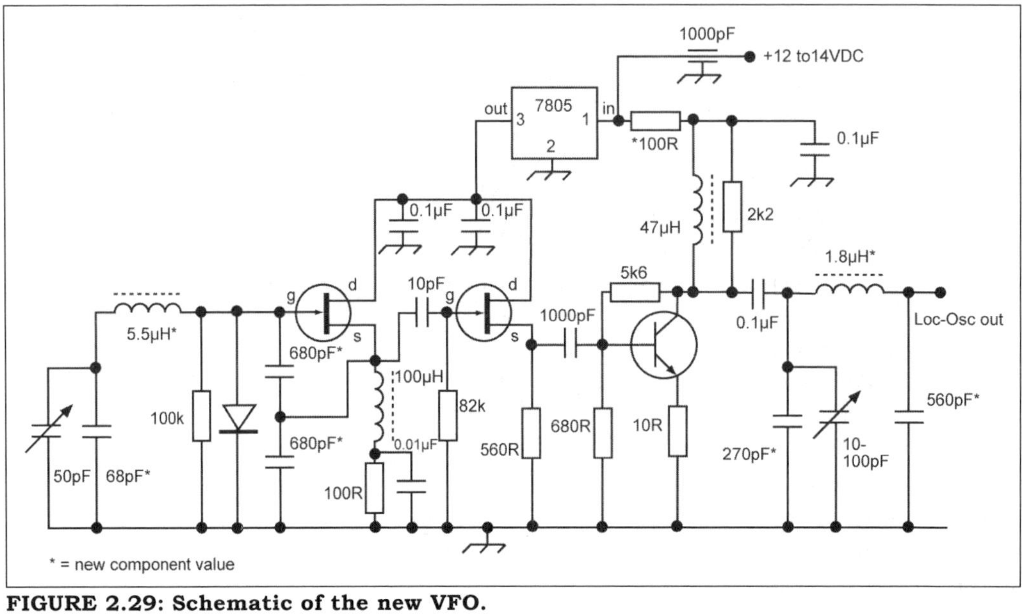

FIGURE 2.29: Schematic of the new VFO.

didn't bother fitting them to my own receiver because I am not too concerned about damaging the crystals in my home made filter. If you are using an expensive IF filter, you would be well advised to use them (do as I say, not as I do). The amplifier is capable of delivering several hundred mW, which might be enough to damage the filter.

LOCAL OSCILLATOR. The local oscillator from the simple DC receiver, described previously, was modified to cover the frequency range required by the new receiver. My receiver has an IF centre frequency of 11.276MHz. As I have chosen to run the local-oscillator on the low side of the IF frequency the local-oscillator frequency is f_IF – f_SIG.

The required local-oscillator frequency span to cover from 3.500MHz to 4.000MHz is 7.776MHz to 7.276MHz.

Note that the VFO tunes backwards with this arrangement. This problem can be solved by running the local oscillator on the high side of the IF. This would probably increase the drift rate of the free-running VFO. I have learned to live with the backwards tuning for now. The schematic of the new VFO is shown in **Figure 2.29**. The inductance of the coil in the main VFO tank circuit has been reduced to 5.5μH. I used a home made inductor wound on a ceramic former. The former is a section from a 1kW heater element as used in the DC RX. 41 turns gave the required inductance. Alternatively, you can use about 32 turns on a T50-2 powdered iron toroid. The 1.8μH inductor in the output network is made from 18 turns of enamelled wire on a T50-2 core.

FIRST TESTS. A few tests were performed on the modules built to date. The VFO output into a 50Ω load was measured with a Marconi mV meter. A pair of 100Ω resistors connected in parallel was used as a load. The output of the VFO was greater than +17dBm across the European 80m band. The measured output was +17.5dBm at the CW end of the band and +19.5dBm at the SSB DX window. The accuracy of the measurements was confirmed with an oscilloscope and also with a diode probe and DC voltmeter.

Some tests with a homebrew signal generator and a few 50Ω resistors showed that the mixer has excellent port to port isolation. The LO port to IF port isolation was measured at 69dB. RF port to IF port isolation was also above 60dB. A home made mixer using 1N4148 diodes and a Mini Circuits SBL-1 mixer module were tested for comparison. The IN4148 mixer had port isolation of around 50dB and the SBL-1 was right on the manufacturer's specification of 60dB. Tests with the signal generator and a precision step attenuator show that the mixer has an input 1dB compression point of +14.5dBm. The tests were performed at 3.579MHz. Even better results could be expected at the SSB end of the band where the VFO output is greater.

The various modules were wired together to do some listening tests. The receiver looks like a rat's nest, but it works well. The low-level AF amplifier from the DC RX project has far too much gain to be used as the first AF amplifier in this receiver. The gain was reduced to a suitable level by simply removing the second stage (see **Figure 2.30**).

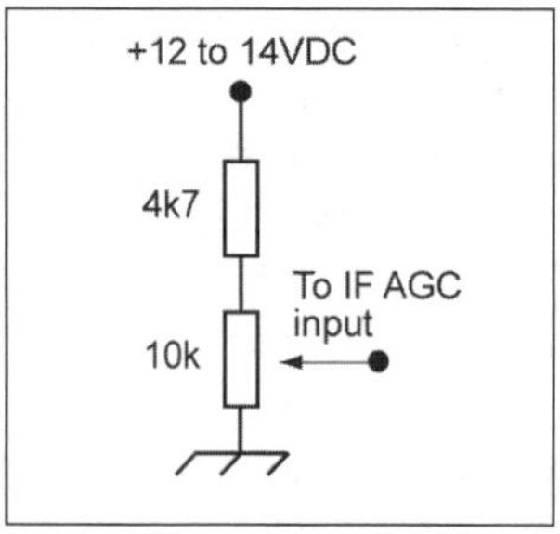

FIGURE 2.30: The gain of the low-level AF amplifier from the DC Rx project was reduced by removing the second stage.

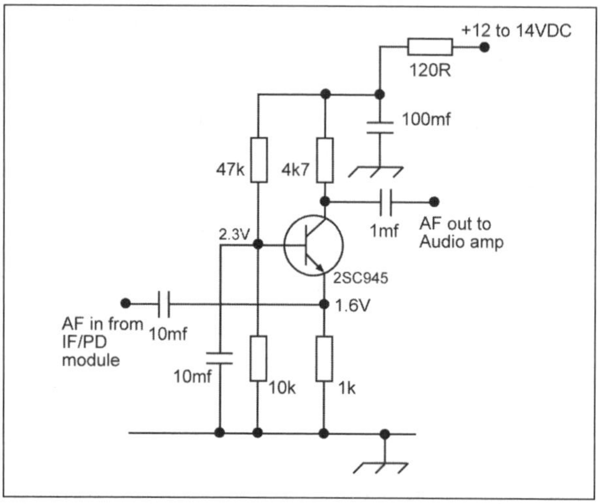

FIGURE 2.31: To do the listening tests, Eamon used a simple manual IF gain control.

Sensitivity is quite good despite the lack of an RF amplifier. Switching between my 80m dipole and a 50Ω dummy load always showed a substantial difference in received noise level, even during the middle of the day when 80m is very quiet. I was able to hear weak DX signals when there were very strong SSB stations at ±3kHz. First indications are that this receiver is quite bomb-proof and compares well with the receiver in my main HF transceiver. To do the listening tests, I used a simple manual IF gain control, see **Figure 2.31** for details. A sample of the received audio can be found at:

http://homepage.eircom.net/~ei9gq/test.mp3

MIXER COMPONENT LIST	
T1, T2	6 turns insulated wire, trifilar wound on FT37-43 or similar ferrite toroid. See text
D1-D4	BAT43 Schottky signal diode

POST-MIXER AMP COMPONENT LIST	
C1,C3	0.01μF
C2,C4,C5,C6	0.1μF
R1	560Ω
R2,R7,R8	1k
R3	3k3
R4	4R7 (4.7Ω).
R5	15Ω
R6	10Ω
R9	120Ω
D5-D8	1N4148 see text
Q1	BFW16A or similar. See text
L1	47μH. See text

Automatic Gain Control System

Designing a good automatic gain control (AGC) system for an SSB or CW receiver has always been one of the trickier aspects of homebrew. A good superhet receiver will have a gain of well over 100dB and an AGC control range of at least 60 to 100dB. To avoid distortion on the first syllable of an SSB speech transmission or the leading edge of a strong CW signal, the AGC must reduce the receiver's gain very quickly. To cope with very strong signals on a quiet band, the AGC system must be able to give a substantial gain reduction within a few milliseconds. The time taken to reduce the receiver gain is known as the AGC attack time. When a strong signal is no longer present, the AGC system will restore the receiver to maximum gain. The time taken to recover full or almost full gain is known as the AGC decay time. The decay time is usually much longer than the attack time.

Most receivers give a gradual gain increase during the AGC decay period. A few receivers use a different system where the gain remains constant for most of the AGC recovery period, then returns very rapidly to full gain. This system is sometimes referred to as 'hang AGC'. My own preference is for a conventional AGC system with a slow decay time of several seconds. For fast break-in CW or contest operators, a shorter decay time might be preferable.

We must decide at this point whether we will use IF derived or audio derived AGC. An IF derived system would measure the level of received signals at the IF frequency. Audio derived systems measure the strength of received signals at audio frequencies. IF derived AGC systems have better dynamic characteristics than audio derived systems. It is easier to get fast attack rates when we are sampling a signal of several MHz. Audio derived systems are usually easier to build. It is easy to arrange an audio derived system so that audio frequency filters are included within the AGC control loop. As an example, consider a receiver that is fitted with an audio notch filter. When the notch filter is used to notch out an interfering tone, the tone can no longer be heard by the operator.

This inaudible signal will still operate an IF derived AGC system and reduce receiver gain. This problem can be avoided in an audio derived system by taking the audio signal for the AGC from one of the stages following the notch filter. This approach must be used with caution. Audio filters with a very narrow bandwidth will limit the AGC attack time. I have built several different AGC systems for this receiver. As usual, the simplest one turned out to be the best.

Figure 2.32 shows a simplified circuit of a typical AGC system. The first amplifier stage increases the voltage of the IF or AF signal to a suitable level for driving a diode rectifier. The DC output from the rectifier is stored in capacitor C. The first amplifier must be capable of driving a relatively low impedance so that C

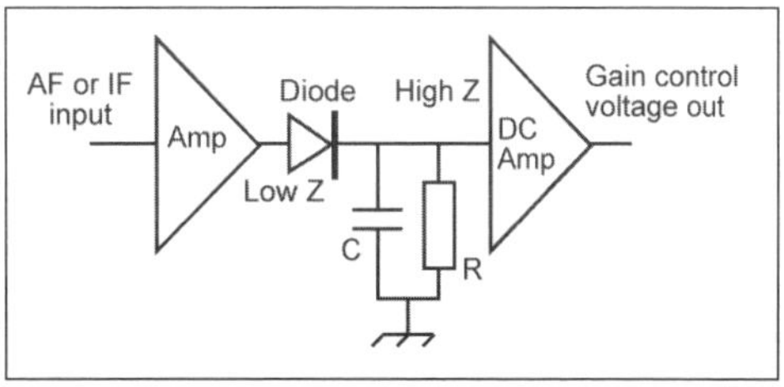

FIGURE 2.32: A simplified circuit of a typical AGC system.

can be charged very quickly. During the AGC decay period, the voltage across C falls slowly as the capacitor is discharged by the current through resistor R. The value of C is quite small, usually no more than a few microfarads. Since the AGC decay rate is determined by the time constant of C and R, the value of R must be quite high, typically a megohm (one million ohms) or more.

The DC amplifier acts as a buffer between this RC network and the IF amplifier and any other circuits that are driven by the AGC system. The DC amplifier must have a very high input impedance so that it will not have a significant effect on the AGC time constants. As well as providing the AGC control voltage, the output from the DC amplifier can be used to drive an S-meter and/or squelch circuit.

THE CIRCUIT. After many hours of experimentation with several different IF and AF derived AGC systems, the circuit in **Figure 2.33** was found to give the best results. This is an audio derived system that produces a negative-going AGC voltage when an audio signal is applied to the input. The AF amplifier stage needs to have a fairly high input impedance so that it will not load the low-level audio stage of the receiver. It must also have a low output impedance. I was planning to use an opamp IC for this stage until I remembered that I already have a perfectly good design for a suitable amplifier. The audio output stage from the amplifier previously built is perfect for the job. The LM386 audio amplifier IC is, after all, just a high-power op-amp with the bias and negative feedback resistors contained within the IC.

A 50kΩ pot is used to set the AGC gain. The value of this pot is not too critical, a 47kΩ pot might be easier to find, anything from 3k to 100kΩ will probably do just as well. I used a BAT43 Schottky diode as the AGC rectifier/detector because it has a lower turn on voltage than standard silicon diodes. The BAT43 is readily available from Maplin and many other sources.

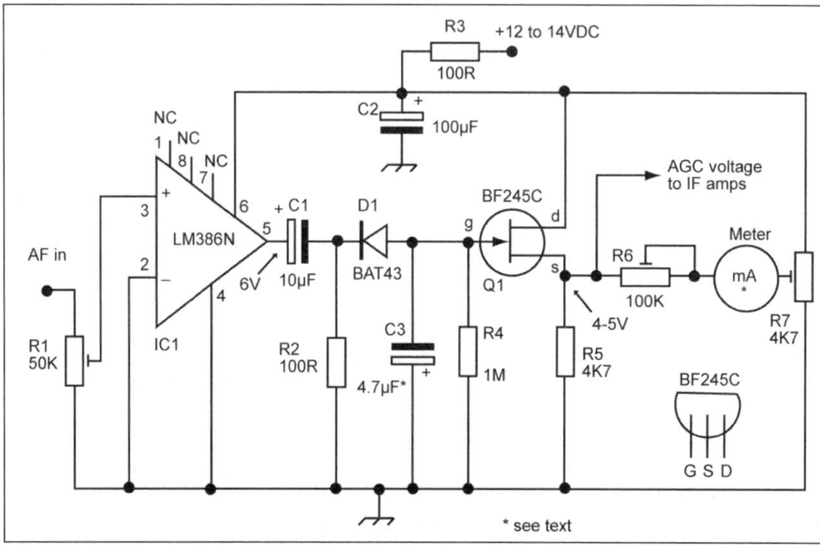

FIGURE 2.33: After experimenting with different IF and AF derived AGC systems, this circuit was found to give the best results.

You could probably use a standard switching diode like the IN4148 or 1N914 instead.

Germanium point-contact diodes should not be used in this circuit because of their relatively high leakage when they are reverse biased. C3 is a tantalum electrolytic capacitor. This component was chosen for its very low leakage resistance and low ESR at audio frequencies. This is one of those rare occasions when the exact choice of component might actually make a real difference. If you decide to use a standard aluminium electrolytic instead, make sure that it is a high quality, low leakage type.

A JFET source-follower is used as the DC amplifier. Because the FET has a very high input resistance, the AGC decay rate is almost entirely dependant on the time constant or C3 and R4. My own personal preference is for a very slow decay rate of several seconds. If you would prefer to have faster AGC, you could use a smaller value for C3. It would be quite easy to provide selectable fast/slow AGC by using a pair of capacitors and a switch on the receiver front panel. R5, R7 and the JFET form a resistance bridge which is used to drive the signal meter. To calibrate the meter, the aerial should be disconnected from the receiver or preferably replaced with a 50Ω resistor. R7 is then used to zero the S-meter. R6 is used to adjust the maximum meter deflection while receiving an extremely strong signal. I used a 500μA meter as the S-meter. Any meter with an FSD current in the range 50 to 500μA should be suitable.

CONSTRUCTION. Like most of the other modules in the receiver, the AGC unit was built on copper-clad PCB using dead-bug point to point wiring. The layout of the circuit is not very critical. You should make sure that the AGC unit and the IF amplifier are both well connected to a common ground. **Photo 2.14** shows the prototype of the receiver. The modules are numbered as follows: **1** RF band-pass-filter, **2** RF frontend (mixer and post mixer amp), **3** crystal ladder filter, **4** IF amplifier, **5** low-level AF amplifier, **6** AGC, **7** audio power amplifier and **8** 12 to 14VDC power distribution.

The DC negative supply is connected to the metal chassis (ground) and the DC positive supply enters the metal enclosure through a 1,000pF feed-through capacitor. The positive supply goes via a large RF choke to the DC power distribution point.

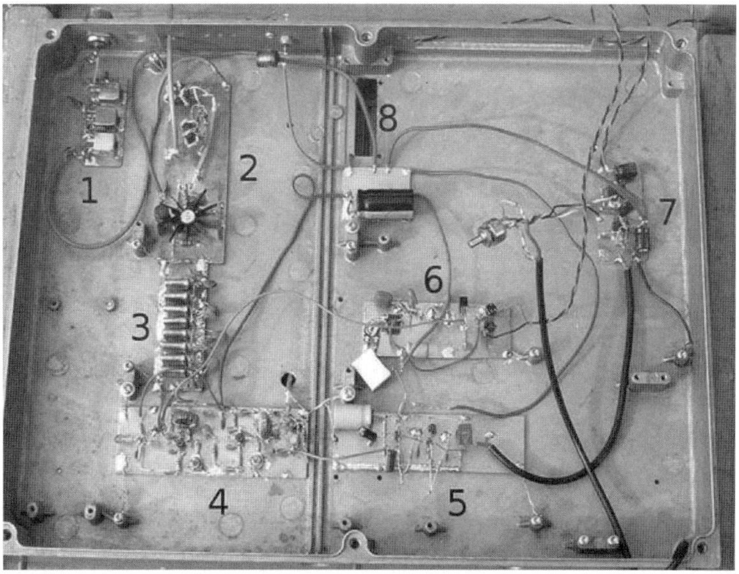

PHOTO 2.14: The prototype of the receiver.

The only component on the DC distribution board is a 1000μF 25V capacitor that is connected between +12V and ground. All parts of the receiver that require a 12-14V DC supply are wired to this point. The VFO unit is in a separate enclosure. The IF carrier oscillator (CIO/BFO) is not visible because it is mounted on the opposite side of the chassis. This brings us to a very important point: The IF carrier unit generates a very strong signal at the receiver's IF frequency. This signal must be kept well away from the input of the IF amplifier stages. The carrier oscillator should be in a well screened enclosure or mounted on the opposite side of the chassis to the IF amplifiers. If you look closely at the photo, you can see the screened co-axial cable that brings the carrier oscillator signal to the product detector via a hole in the chassis.

AUDIO LPF. To improve the receiver's signal-to-noise ratio, I added a very simple low pass filter to the low level audio amplifier stage. The filter is a Sallen-Key type active filter using a pair of transistors and a few resistors and capacitors. The component values in the schematic will give a -3dB cut-off frequency of about 2.1kHz which is suitable for SSB reception.

For narrow bandwidth modes like CW, PSK and RTTY, the value of the 6.8nF capacitors can be increased to 100nF. As well as reducing the bandwidth, this will increase the Q of the filter so that the response is more like that of a band pass filter. If you want to have selectable SSB/CW bandwidth, you can use a DBDT (double pole, double throw) switch to select the capacitors. The revised low level amp unit is shown in **Figure 2.34**. The common-base low level amplifier Q1 is identical to the previous versions except for the use of BC548 NPN transistors instead of the 2SC945. LTSpice/SwCAD **[20]** software was used to simulate the behaviour of the active filter. Amplitude vs frequency plots for the SSB and CW versions are shown in **Figure 2.35**.

CALIBRATION AND ALIGNMENT. Before calibrating the AGC and S-meter, you should check that the RF band pass filter and IF amplifiers

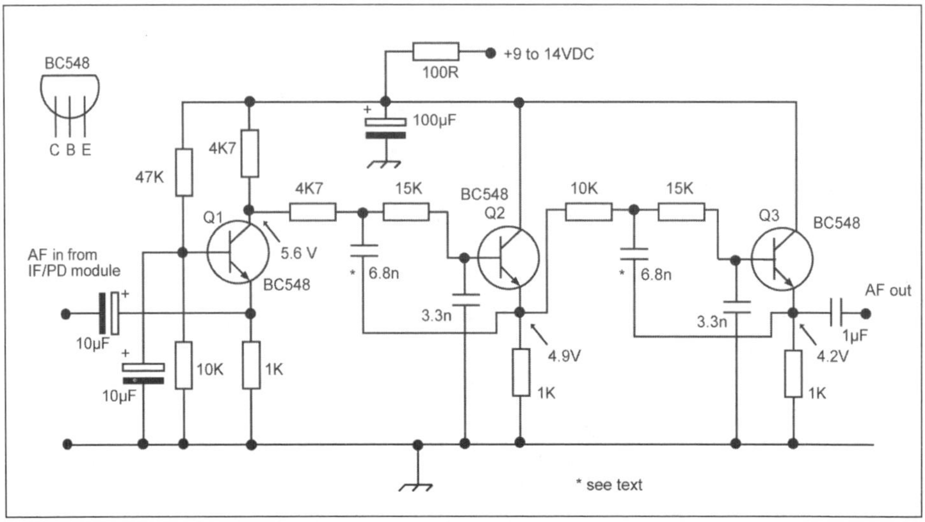

FIGURE 2.34: The revised low level amp unit.

are properly tuned. Set R1 on the AGC unit to the minimum gain position. This will disable the AGC and the receiver will operate at maximum gain.

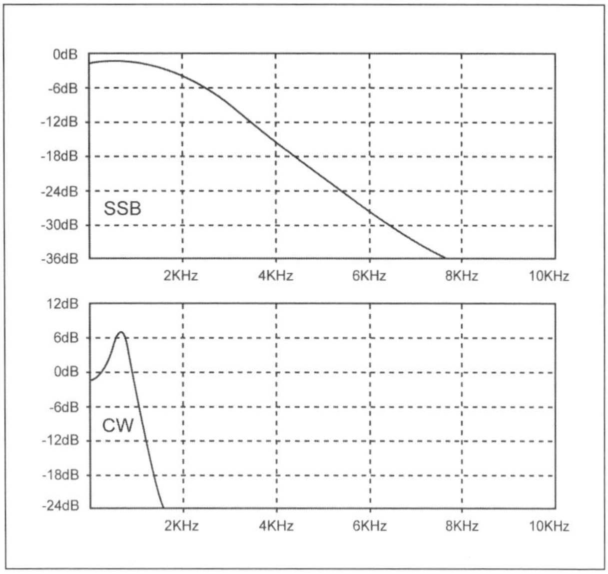

FIGURE 2.35: Amplitude vs frequency plots for the SSB and CW active filters.

Adjust R7 on the AGC unit to zero the meter. Use a signal generator if you have one. If you don't, just find a signal near the middle of the band and adjust the RF BPF for best signal strength. Check the sensitivity at the upper and lower band edges and re-tune until you have a reasonably flat response across the entire band. Next, tune the receiver to a weak signal anywhere in the band and adjust the IF amplifier tuning capacitors for best signal. Take your time at this stage, go over the adjustments several times. Once the RF and IF stages are aligned correctly, you should resist the urge to go back and tweak them again. Once the filters are aligned correctly, no amount of tweaking will make any improvement. Tune the receiver to a very strong signal. You should try to find the strongest signal you are likely to encounter in normal use. Some of the commercial, broadcast or weather FAX stations should provide a source of multimillivolt signals during the hours of darkness. A signal this strong will completely overload the IF and PD stages.

Adjust R1 on the AGC unit until the gain has decreased to a point where the receiver is no longer overloaded. Adjust R6 until the meter is reading just short of full scale.

There is some interaction between the adjustment of R6 and the meter zero pot (R7). You will need to repeat this procedure several times until the meter gives reasonable readings. I have found that I can calibrate the meter so that a moderately strong signal of about 50μV or S9 is at the middle of the scale and even the most powerful signals on the band will not drive the meter to the end stop.

This AGC circuit has behaved very well in the simple superhet. It copes well with all signal levels under varying band conditions. It even survived the ultimate test of receiving a very strong local station, running legal limit power output under quiet daytime conditions. As usual, an audio sample can be found at:

http://homepage.eircom.net/~ei9gq/biggun.mp3

To complete the simple superhet project, we need a well calibrated frequency display. We could use a mechanical/analogue dial or we could use one of the nice digital frequency counter/display units that are available in kit form. I would prefer to build a frequency counter from scratch.

AGC COMPONENT LIST	
R1	50kΩ Pot. 47kΩ is close enough
R2, R3	100Ω
R4	1MΩ
R5	4k7
R6	100kΩ preset pot. See text
R7	4k7 (4.7kΩ) preset pot
C1	10μF electrolytic
C2	100μF electrolytic
C3	4.7μF Tantalum or other low leakage electrolytic. See text
D1	BAT43
IC1	LM386N
Q1	BF245C or similar JFET
Meter	micro-ammeter. See text

References

[10] The TL071 data sheet is available from the STMicroelectronics website: http://www.st.com If you are searching the web for data, note that the device name is TL071 and not TLO71. The third character is the number zero. The PDF file name it tl071.pdf

[11] The LM386N data sheet is available from National semiconductor. The file name is: LM386.pdf http://www.national.com

Both the above data sheets are also available from Maplin's website: http://www.maplin.co.uk

[12] A Unified Approach to the Design of Crystal Ladder Filters by Wes Hayward W7ZOI. QST, May 1982.

[13] ViPEC Electrical network analysis for Linux and other UNIX based systems. http://vipec.sourceforge.net/ RFSim99 for Windows has similar functionality.

[14] Xanalyser Audio spectrum analysis for Linux. http://arvin.schnell-web.net/xanalyser/

[15] Windows users could try Spectrogram instead. www.visualizationsoftware.com/gram.html

[16] Handbook of Filter Synthesis by Anatol I Zverev. Wiley 1967

[17] www.amidoncorp.com

[18] www.w8ji.com/receivers.htm

[19] www.st.com search for 'BAT43'

[20] www.linear.com/company/software.jsp

3

Frequency Measurement

A MICROCONTROLLER is a miniature computer on a single chip. A typical device will have the CPU core, clock oscillator, memory, I/O ports and counter/timer all integrated into a single IC. Many microcontrollers have ADC inputs, PWM outputs, serial communications and a host of other features. You should not be intimidated by the apparent complexity of microcontroller circuits. In fact, most microcontroller circuits are extremely simple. For a typical amateur application like a Morse keyer or repeater controller, you will only need the microcontroller IC, a crystal for the clock oscillator and a few discrete components. If exact timing is not required, the crystal can be replaced by a simple RC network. **Figure 3.1** shows a schematic of a microcontroller based iambic keyer **[21]**.

The PIC micro-controllers made by Microchip Technology are very popular with home constructors. Flash EEPROM devices like the PIC16F84 and PIC16F628 are particularly useful because they are electrically erasable and reprogrammable. We will be using the PIC16F628. The schematic of the programmer is shown in **Figure 3.2**. If you don't want to go to the trouble of etching the PCB, the circuit can easily be built on a small strip of Veroboard. The circuit is based on the JDM design by Jens Madsen.

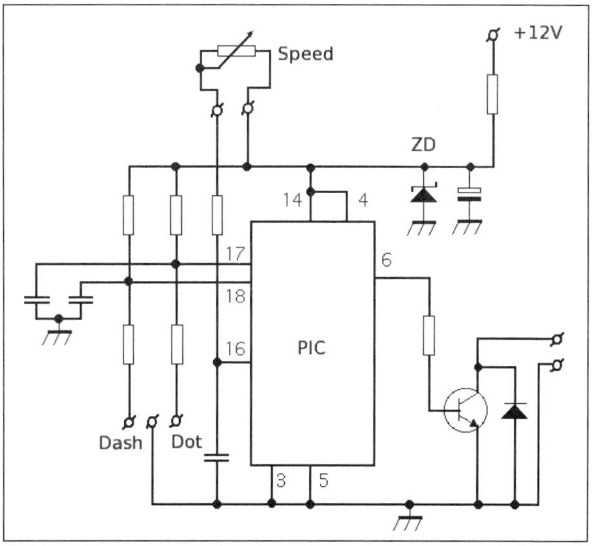

Figure 3.1: A schematic of a microcontroller based iambic keyer.

PIC microcontrollers are based on a RISC (reduced instruction set computer) architecture. This means that they have a very small and simple

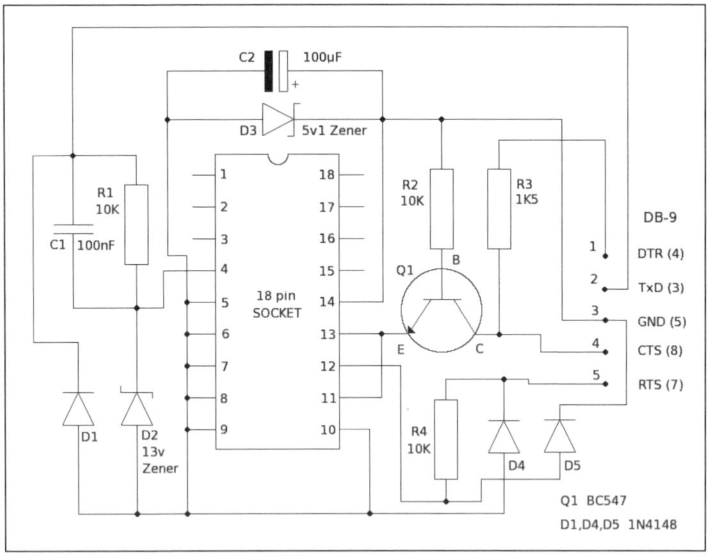

Figure 3.2: The schematic of the PIC programmer.

instruction set. The PIC16F628 has a total of 35 instructions. Each individual instruction is very simple and easy to learn. A typical example of a PIC instruction in assembly language is 'movlw'. This instruction takes a literal value like 2 or 99 and places it in the w register. The w (working) register is a special register that is used for simple arithmetic and logical operations. **Box 3.1** contains an example of a very simple assembly language program.

This program takes the number 2 and stores it in one of the PIC's general purpose file registers (a fancy name for RAM) at address 32 hexadecimal. It then places the number 1 in the w register. The final instruction addwf (add working register to file register) adds the value that is currently in the working register to the value stored in the specified file register. The ',f' in the last instruction specifies that the result of the addition should be stored in the file register and not in the w register. The actual machine code that is executed by the PIC is a sequence of 14-bit binary numbers. The code for 'movlw 2' is: 11000000000010. Long sequences of binary numbers are very difficult for humans to read. Assembly language programs use easily recognisable words called mnemonics to represent the binary machine code instructions. This human-readable code is known as the source code. The binary code produced by the assembler is known as the object code. The main job of the assembler is to translate the source code into binary machine code.

The PIC assembler has many other useful features including the ability to assign numeric values to text labels. Binary numbers are usually written in hexadecimal (base 16) notation. Hexadecimal uses a single digit to represent four binary bits. Decimal numbers allow a

```
movlw    2          ;place first number in working register
movwf    0x32       ;store first number at address 32 HEX
movlw    1          ;place second number in working register
addwf    0x32,f     ;add the two numbers together
```

Box 3.1. An example of a very simple assembly language program.

maximum value of 9 in any one digit. Four bits of binary allow a maximum range of 16 (4^2) values from 0 to 15. Hexadecimal uses the letters A to F to represent the numbers 10 to 15. An 8-bit binary number is known as a byte and has a range of 256 values from 0 to 255 (00 to FF hex). Numbers greater than 255 can be handled by using multiple bytes. Numbers written in hexadecimal are usually preceded by '0x' to tell the assembler that the following number is a hex number and not decimal, binary or octal (base 8).

Despite the fact that there are many different CPU architectures, each with its own unique instruction set, most assemblers comply with a similar set of rules and conventions. Assembly language programs usually have labels representing program memory addresses as the first word of each line. This label is followed by some 'white space' in the form of one or more spaces or tabs. The next word on the line is the mnemonic for the instruction. The instruction is often referred to as the opcode or operator.

Many instructions require numeric data such as addresses or literal values. This data is contained in the third word or 'field' of the line and is referred to as the operand. For instructions that require more than one operand, a comma is used to separate the operands. For instructions that don't require an operand, the third field should be left blank. The PIC assembler will ignore any text that is preceded by a semicolon. Text comments are usually written in the fourth or final field of each line. **Box 3.2** shows a typical line of code.

It is not possible to fit a complete PIC programming and assembly language tutorial into a few pages. The MPASM assembler users guide **[22]** alone runs to 246 pages. The PIC16F628 data sheet **[23]** is 180 pages. The data sheets contain a lot of useful information so you should take the time to study them carefully. There are also many excellent PIC tutorials on the web.

SOFTWARE TOOLS. We will need to acquire a few software tools before we begin. All of the software we need is available free of charge. All of the software mentioned here was tested on a humble Pentium II 333MHz and a more modern Athlon-XP.

EDITOR. The first thing we will need is a text editor for writing the source code. The editor can be part of an IDE (integrated development environment) like MPLAB or you can use a standard text editor. Notepad for Windows is not a good programmers' editor because it can not display line numbers and it can't cope with Unix or Mac text files. A programmers' editor like notepad++ would be a better choice. I use an

```
label     opcode    operand/s     ;Comments go here
flash     bsf       PORTA,LED     ;Turn LED on
```

Box 3.2. A typical line of assembler code.

IDE called Pikdev on my Linux PC. **Photo 3.1** shows the Pikdev editor window with the built-in PIC programmer window in the foreground.

ASSEMBLER. Microchip's assembler for Windows is included with MPLAB. The assembler component can also be downloaded separately as MPASMWIN. Pikdev for Linux uses the Gpasm assembler from the Gputils project.

PROGRAMMING. Simple serial port programmers like the JDM are very well supported by most PIC programming software. I have used the programmer in Figure 3.2 to program PIC16F628, PIC16F84 and PIC16C84 microcontrollers. The programmer has been tested and found to work perfectly with Pikdev and Picprog under Linux. It also works perfectly with IC-Prog, and WinPic under Windows XP. This programmer will probably work with the PIC16F627 and some other 18-pin PICs. See the Picprog and IC-Prog documentation for a full list of supported PIC devices. The programmer connects to a standard RS232 serial port. The pin numbers shown are for 9-pin serial port connectors. The programmer 'steals' its power supply voltages from the serial port. Most desktop PC serial ports can deliver ±12V. Some laptop computers have a reduced serial port voltage of ±5V and will not work with this programmer. Note that most PIC programming software requires a 'real' serial port and will not work with a USB-to-serial converter.

SIMULATORS. Very few PIC programs will work perfectly first time. Even the most skilled programmers will make mistakes. A PIC simulator is a software tool that allows you to run your PIC code on a standard PC. The big advantage of running your code on this 'virtual PIC' is that you can analyse your program step by step and see the effect of every instruction on the PIC's registers and I/O pins. A simulator is a very useful tool for debugging and fine-tuning your code. MPLAB for Windows has a built in simulator. I use a simulator called Gpsim (**Photo 3.2**) on my Linux PC.

Gpsim is incredibly fast, even on the old Pentium II machine. It can run PIC code as fast as a real PIC with a 10MHz clock!

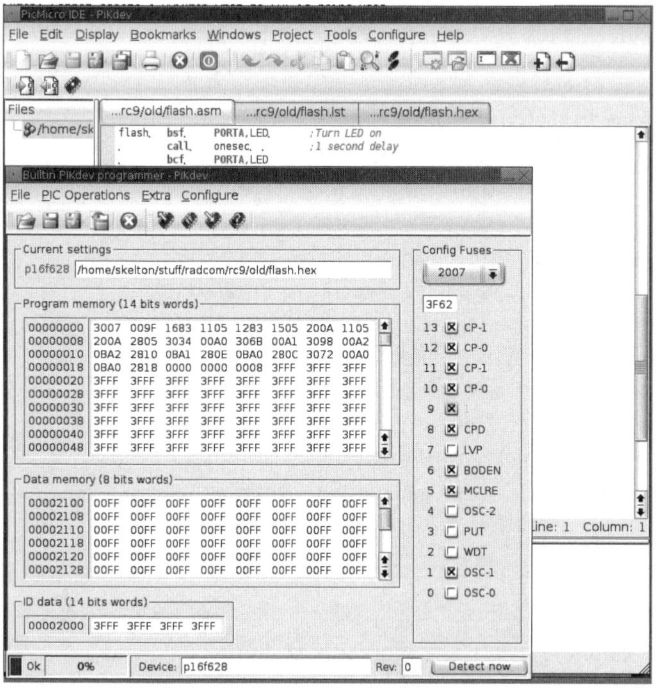

Photo 3.1: The Pikdev editor window with the built-in PIC programmer window in the foreground.

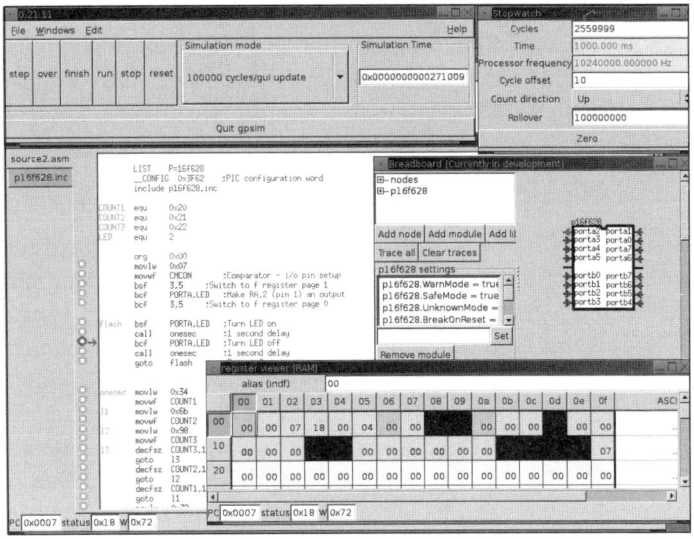

Photo 3.2: Screenshot of Gpsim.

FIRST PROJECT. It is a well established tradition in the PIC programming world that novice PIC developers should start by learning to program a PIC to carry out the simple task of flashing an LED. This seemingly trivial task will show you how to configure and use the PIC I/O ports and gets to the very core of what PICs are all about. The schematic of our test circuit is shown in **Figure 3.3**. The PIC needs a supply voltage of around 5V. If you don't have a 5V bench supply, you can use a 7805 regulator (as shown within the dashed lines) with a 12V supply or a 9V battery. Check the supply voltage of your test circuit before you plug the PIC in. The absolute maximum supply voltage rating of the PIC16F628 is 6.5V. Exceeding this limit will probably destroy the PIC. I built two versions of the test circuit, one on a solderless breadboard

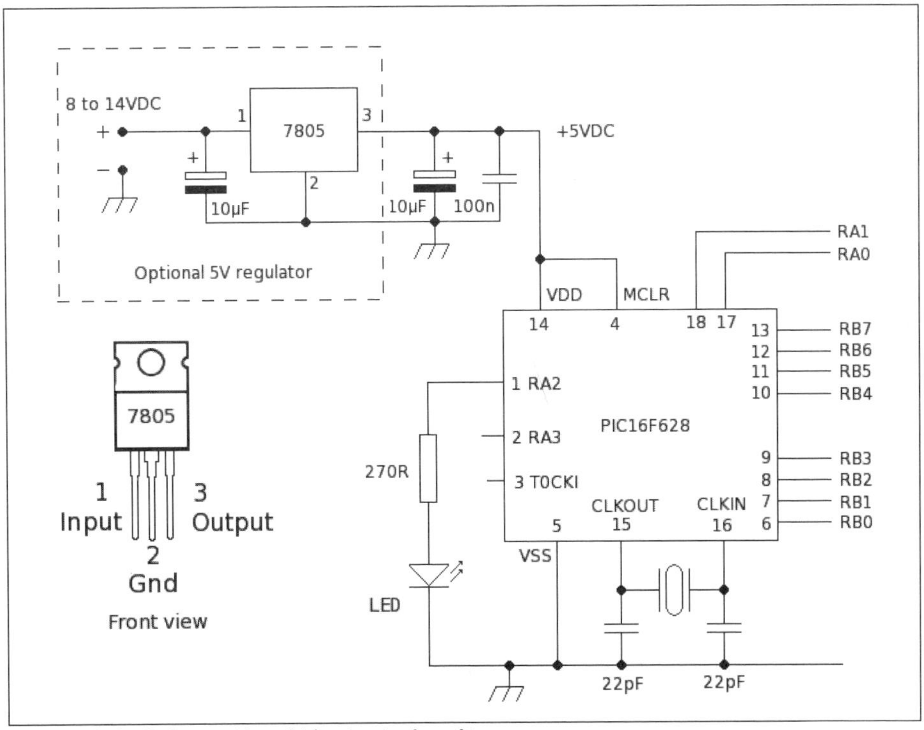

Figure 3.3: Schematic of the test circuit.

and the other on Veroboard. **Photo 3.3** shows the breadboard version; **Photo 3.4** shows the Veroboard version. I used a 10.240MHz crystal for the clock oscillator. The PIC is a PIC16F628-20/P, which is the 20MHz version in a plastic DIL package. The PIC16F628A should work just as well. If you can't find a 20MHz PIC in your junkbox, you could try using a 4MHz PIC like the PIC16F628-04/P. I have never encountered a 4MHz PIC that wouldn't work reliably up to at least 16MHz.

We will now work through the process of writing the program, generating the PIC code with an assembler, programming the PIC chip and testing it in our simple test circuit. A typical PIC program is much larger and more complicated than this simple example. Our program is so small and simple that I can include a full listing of the source code. **Figure 3.4** shows the source code for our program. The coloured syntax highlighting and line numbers were added automatically by the Pikdev editor. You should copy the source code line by line (without the line numbers) and save it as a file called led.asm. You should consult the PIC data sheet to get a detailed description of each instruction as you type it in.

We will now examine the source code of our program to see if we can figure out what it does and how it does it. Line 1 contains an assembler directive. This is not a PIC instruction, it is a command to the assembler that tells it to generate object code for the PIC16F628 and not for the many other PICs that are supported by the same assembler. Line 2 has another directive that defines the value in the PIC configuration word at address 2007 hex. The value 3F62 hex configures the PIC to use a high-speed crystal oscillator, WDT off and power up timer on. See the PIC data sheet for full details. Line 3 tells the assembler to include the standard header file for the PIC16F628. This file contains many standard definitions like PORTA, PORTB, CMCON and so on.

Line 5, 6 and 7 define the values of three constants. These are the addresses of the three general purpose file registers that are used in the delay loop subroutine. Line 8 defines the PORTA bit that drives the LED. Line 10 tells the assembler that the following instruction should

Photo 3.3: Breadboard version of the test circuit.

Photo 3.4: Veroboard version of the test circuit.

```
1            LIST     P=16f628
2            __CONFIG  0x3F62          ;PIC configuration word
3            include p16f628.inc
4
5   COUNT1   equ      0x20
6   COUNT2   equ      0x21
7   COUNT3   equ      0x22
8   LED      equ      2
9
10           org      0x00
11           movlw    0x07
12           movwf    CMCON            ;Comparator - i/o pin setup
13           bsf      3,5              ;Switch to f register page 1
14           bcf      PORTA,LED        ;Make RA,2 (pin 1) an output
15           bcf      3,5              ;Switch to f register page 0
16
17  flash    bsf      PORTA,LED        ;Turn LED on
18           call     onesec           ;1 second delay
19           bcf      PORTA,LED        ;Turn LED off
20           call     onesec           ;1 second delay
21           goto     flash            ;Repeat forever.
22
23
24  onesec   movlw    0x34             ;1 second delay loop
25           movwf    COUNT1           ;Crystal freq -
26  l1       movlw    0x6b             ;10.240000MHz
27           movwf    COUNT2
28  l2       movlw    0x98
29           movwf    COUNT3
30  l3       decfsz   COUNT3,f
31           goto     l3
32           decfsz   COUNT2,f
33           goto     l2
34           decfsz   COUNT1,f
35           goto     l1
36           movlw    0x72
37           movwf    COUNT1
38  l4       decfsz   COUNT1,f
39           goto     l4
40           nop
41           nop
42           return
43
44           end
```

Figure 3.4: The source code.

be placed in memory address 0. The instruction at this address is the first instruction that will be executed when the PIC is reset. Our program begins executing at this point. The PIC instructions in the next few lines are used to configure the PIC I/O pins. Line 11 contains the familiar movlw instruction that places the value 7 in the working register. Line 12 stores this value in a special function register that controls the behaviour of the port-A I/O pins. Line 13, 14 and 15 configures the port-A data-direction register so that pin 1 (PORTA, LED) is an output. Most PIC programs will begin by configuring the PIC hardware. You will need to configure I/O pins to act as input or outputs, set up the initial state of counters, timers, prescaler, interrupts etc.

The main part of our program is contained in lines 17 to 21. Line 17 sets the appropriate bit of the port A data latch high to turn on the LED. The instruction in line 18 calls a subroutine that implements a one second delay loop. When a subroutine is called in this way, the address of the current instruction is stored in a special register called the stack. Program execution then jumps to the start of the called subroutine, in this case, the address pointed to by the label 'onesec'. Execution of the subroutine will continue from this new address until the return instruction at the end of the subroutine. When the return instruction is executed, the address that was saved in the stack is recovered and program execution continues from the next address after the call instruction. This is an efficient way of coding because it allows us to write a block of code once and re-use it several times in the same program. The bcf (bit clear F) instruction at line 19 clears the PORTA, LED bit, which turns off the LED. Line 20 calls the one second delay loop subroutine again. Line 21 causes program execution to jump to the address of the text label 'flash', which causes the LED to flash forever in an endless loop.

Instructions that cause program execution to jump to a new address instead of executing the instruction at the next sequential address are called program branches. The goto instruction is the simplest type of program branch because it will always unconditionally jump to the new address. Some other PIC instructions will only cause conditional branches. This means that a branch will only happen if certain conditions exist. One example of a conditional branch can occur when we use the 'decfsz' (decrement F skip if zero) instruction. This instruction will decrement (decrease by 1) the value in the specified file register. If the new value in the decremented register is not zero, the next instruction will be executed as normal. If the decremented register contains the value zero, the next instruction will be skipped and execution will continue from two instructions beyond the decfsz instruction. The decfsz instruction is used several times within the one second delay loop. To implement a simple delay loop, we can load a file register with a number greater than zero and use the decfsz instruction to decrement this number repeatedly until the register contains zero. decfsz will then cause a branch that breaks the loop and ends the delay. The code is **Box 3.3** is an example of a short delay loop.

To achieve precise timing, you must know the time required to execute each instruction. The instruction cycle time of the PIC is exactly four times as long as the clock period. With a 4MHz clock, this is exactly 1 microsecond. For a 10.24MHz clock the instruction time is (1/10.24)*4 = 0.390625 microseconds.

To calculate the exact time delay, we must work out the number of instruction cycles that will be executed. Most PIC instructions will execute in one cycle. Instructions that cause program branches take two cycles. The first two instructions take one cycle each, the decfsz instruction takes one cycle unless it encounters a zero and causes a branch. The goto instruction always causes a branch, so it always takes two cycles. This means that the third and fourth instructions form a loop that will execute over and over again until the file register with the label count1 contains the value zero. count1 is one of the PIC's general purpose file registers that we are using as a place to store the count. We know that each execution of the loop will take three cycles, one for the decfsz and two for the goto, except for the final loop when the goto will not be executed. The count register has been loaded with 255 decimal (FF hex) so we know that the main loop will take 3 cycles for each of the first 254 iterations of the loop and a further 2 cycles for the final jump out of the loop. A few pokes at the keypad of my pocket calculator tells me that 1 + 1 + (254 x 3) + 2 = 766 cycles. As each cycle takes 0.390625 microseconds, the total delay time is just over 299.218 microseconds. I assembled the program and ran it on the Gpsim simulator. To my great surprise, the StopWatch window in Gpsim says that the loop takes 766 cycles and 299.22 microseconds.

As 0xFF (255) is the largest number that we can store in a file register, this is almost the longest delay we can implement using such a simple delay loop. You can make it 3 cycles longer by loading count1 with 0 instead of FF. Can you see why? Try running it on the simulator. Long delay times of seconds, minutes or even days can be achieved by using nested loops (loops within loops) as I have done in the onesec subroutine. The assembler directive on line 44 tells the assembler that we have reached the end of the program.

When you have finished typing the source code, you should invoke the assembler by pressing the assemble button in the IDE or selecting Quickbuild from the Project menu. If you are fortunate, the code will assemble without errors or warnings. If not, you should read the error messages and re-edit the source code until the problem is fixed. The source code for this project assembles without errors under Pikdev and MPLAB. The assembler should generate several new files. One of these files is the PIC object code. The file should be called led.hex or something similar.

All that remains for us to do is to 'burn' the program from the hex file into the PIC. The procedure is pretty much the same for Pikdev, IC-Prog and WinPic. Connect the PIC programmer to a PC serial port. This is

```
        count1   equ 0x21
        movlw    0xff          ;FF hex = 255 dec
        movwf    count1        ;count register
loop    decfsz   count1,f      ;decrease count
        goto     loop          ;and again until zero
        nop                    ;don't do anything
```

Box 3.3. Code for a short delay loop.

usually COM1 or COM2 under Windows. Place the PIC in the programmer's IC socket and start up the PIC burning software. I will describe the exact procedure for IC-Prog v1.05 under Windows XP. When I started IC-Prog the status bar at the bottom of the IC-prog window read: JDM Programmer on Com1 Device: PIC 16F628 (100). Obviously, IC-Prog had automatically detected the programmer and configured itself to use COM1 and the correct settings for the JDM programmer. It then read the PIC's ID code. Clever stuff!

After a quick check of the 'Settings:Hardware' menu to confirm that the hardware configuration was correct, I selected 'PIC 16F628' from the device selection listbox at the top right hand side of the window. I opened the led.hex file using the **'File:Open'** menu and clicked the 'Program All' button in the menu bar. It took about 45 seconds to program and verify the chip. This seems like a long time to program just 30 memory locations. I suspect that IC-Prog programs all 2048 program memory locations of the PIC. When the program/verify operation was finished, IC-Prog reported 'Device Successfully Verified' so it seemed that all had gone well. Programming the PIC with WinPic under Windows and Pikdev under Linux was equally painless and quite a bit faster. Programming and verification took 0.81 seconds with Picprog under Linux.

TESTING. When I put the PIC in the test circuit and powered it up, I was greeted by the cheerful sight of a flashing LED. The flashing rate is exactly two seconds or if you prefer, 5,120,000 cycles with a 10.240000MHz clock oscillator. Don't take my word for it, fire up the PIC simulator and check for yourself. You might be unimpressed by this high-precision LED flasher. However, the programming techniques we have learned can now be applied to more useful projects. The PIC development software and device programming hardware has been tried and tested and we are ready to build more advanced projects in the future.

PIC-based Frequency Counter

The next project is to build a simple PIC-based frequency counter.(**Photo 3.5**). The frequency counter hardware is derived from the circuit used for the LED flasher project. A PIC16F628 microcontroller is used to do the counting. The frequency is displayed on a two-line by 16-character LCD display module.

A frequency counter is used to measure the number of accumulated events over a given period of time. The events might be cars crossing a bridge, revolutions of a rotating shaft or cycles of alternating current. Most electronic counters are designed to measure pulses of electrical voltage.

Before the arrival of programmable devices like the PIC, most frequency counters were based on logic ICs. **Figure 3.5** shows a block diagram of such a counter. A crystal oscillator is used as a timebase for the counter gate and a chain of BCD (binary coded decimal) counters is used to count the input pulses. Each BCD counter has a separate BCD to 7-segment driver IC to drive the LED display digits. The counter hardware is quite elaborate but it's method of counting is simple. At the start of each counting period, a reset pulse is applied to the BCD counters so that the count is reset to zero or some other predetermined value. The counter gate is then opened so that pulses at the counter input will increment the value in the counter chain. The BCD counter for the first digit can only count from 0 to 9. The BCD counters are arranged in a cascade so that a carry pulse from the first digit will increment the counter for the second digit. Second digit carry pulses will increment the third digit and so on. When the count period is over, the count is frozen by closing the counter gate and an enable pulse latches the result of the count onto the outputs of the seven segment LED drivers. The counters are then reset and the whole process begins again. The timing of the reset, gate and enable pulses is controlled by a very stable oscillator. The count accuracy depends on the frequency accuracy and stability of the timebase oscillator. The count resolution is dependant on the length of time that the gate is open. Dedicated frequency counter ICs like the ICM7216 combine all of the above blocks into a single chip.

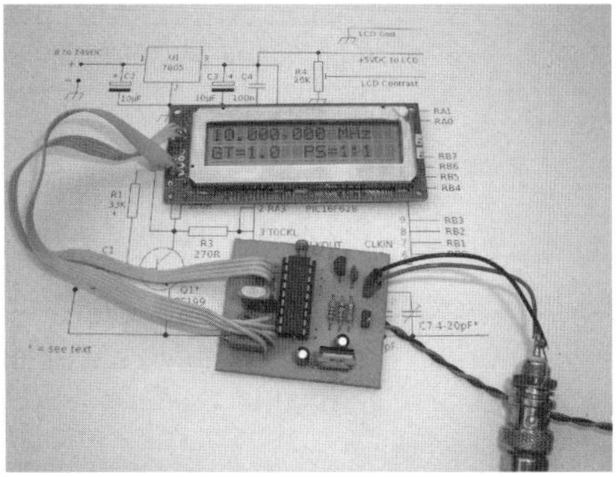

Photo 3.5: The completed PIC-based frequency counter.

THE PIC MICRO. The PIC16F628 is an 8-bit microcontroller with up to 16 user-configurable I/O pins, 224 bytes of RAM, 128 bytes of data EEPROM, and 2,048 x 14 bit words of program memory. The data EEPROM and program memory are electrically erasable and reprogrammable. This means that you can reprogram the PIC thousands of times. The PIC16F628 has several different modes for its main clock oscillator, including: a built-in crystal oscillator; an RC oscillator that uses an external resistor and capacitor; and a built-in RC oscillator mode that needs no external components and leaves the oscillator in and out pins free for use as I/O port pins. The PIC can also take its clock signal from an external source. We will use one of the PIC16F628's three counter/timer modules to count the frequency of a signal applied to the T0CKI (timer0 clock input) pin. Timer0 is an 8-bit register, which can be configured to count pulses from the PIC's internal clock oscillator or

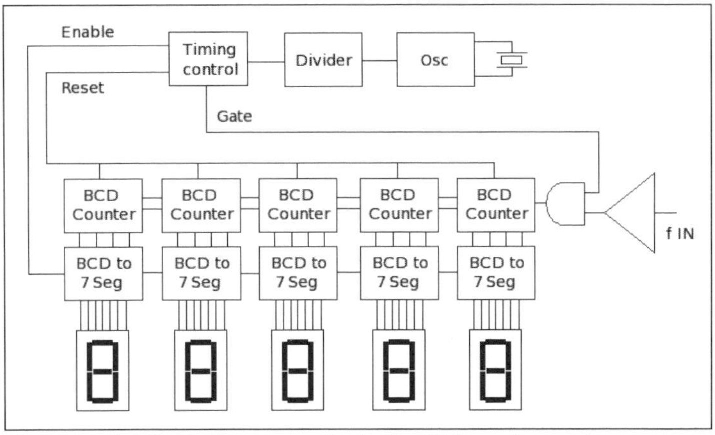

Figure 3.5: Block diagram of a logic-IC-based counter.

from an external clock signal. As the counter is only 8 bits wide, it has a maximum count of 255. This would limit us to a maximum frequency of 255Hz at 1Hz resolution or 255kHz at 1kHz resolution. Fortunately, the PIC has a programmable 8-bit prescaler that can be used to divide the T0CKI signal by 256. The register associated with Timer0 is a normal file register easily read by a simple PIC instruction. The value in the prescaler cannot be read directly. However, we can recover the value by using a trick described in Microchip Application Note AN592 **[24]**. Using the prescaler, increases the maximum count to 16 bits. This allows a maximum frequency of 65.535kHz at 1Hz resolution or 65.535MHz at 1kHz resolution, a big improvement, but not exactly state-of-the-art.

When Timer0 overflows, the T0IF (timer 0 interrupt flag) bit is automatically set in one of the PIC's special function registers. The PIC can be configured so that a Timer0 overflow will cause a hardware interrupt. If timer interrupts are disabled, the overflow flag can simply be polled (tested periodically) to see if an overflow has occurred. Just testing the overflow flag once at the end of the count period would add one more bit to the count and effectively double the frequency range of the counter. There is one big problem with this approach; checking the overflow flag at the end of the count period will tell us if there has been a timer overflow. However, we have no way of telling how many times Timer0 has overflowed. We can overcome this problem by checking the overflow flag frequently during the count period. As soon as an overflow is detected, we can clear the overflow flag again and increment a register that keeps count of the number of timer overflows. This increases the count to a very respectable 24 bits or 16.77MHz at 1Hz resolution or 167.7MHz at 10Hz resolution. Even higher counts can be achieved by using more than one register to keep track of timer overflows.

HOW IT WORKS. Because most of the functions of the frequency counter are performed within the PIC chip, the circuit hardware for our homebrew counter is extremely simple. All we need is the PIC, the LCD display module, a 5V regulator, one transistor and a small handful of passive components. **Figure 3.6** shows a simplified block diagram of the Timer0 input circuit. Note that one of the PIC's I/O pins (RA3) is connected directly to the timer input pin (T0CKI). RA3 is configured as an input during the count (gate open) period, so it has little or no effect on the operation of the counter. We will now look at the sequence of events before, during and immediately after the count (gate open) period. The counter gate is closed by making RA3 on output and using it to clamp the Timer0 input pin low. This prevents any signals from reaching the prescaler input. The registers that contained the result of the last count are cleared to zero using the CLRF instruction. The TMR0 register is also cleared. Note that clearing TMR0 automatically clears the prescaler to zero. Now that the count registers are reset, we can begin the count. The counter gate is opened

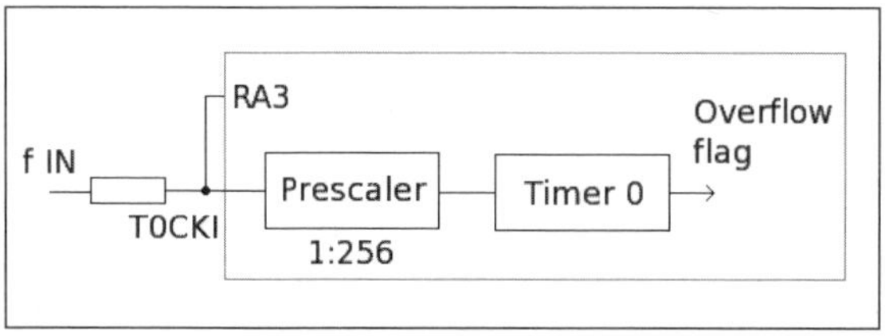

Figure 3.6: A simplified block diagram of the Timer0 input circuit.

by simply making RA3 an input. Once the Timer0 input pin is no longer clamped to ground, any signals at the frequency counter input will be amplified by Q1 and then fed to the Timer0 input pin of the PIC (pin 3). The gate time of the counter is determined by a software delay loop in the PIC program. I have chosen to use a one second delay so that the count resolution will be 1Hz.

During the counter gate period, we need to check the Timer0 overflow flag frequently so that we can keep an accurate count of timer overflows. The prescaler input is rated for a maximum frequency of 50MHz. We can safely assume that the prescaler input of the current generation of PICs will never have to deal with frequencies of greater than 100MHz. TMR0 will overflow every 65,536 cycles of the prescaler input frequency. Since our worst case signal at a frequency of 100MHz has a period of 0.01 microseconds, it is possible that a TMR0 overflow could happen every 655 microseconds. This is quite a long time in the high-speed world of a PIC program. We have plenty of time to check the overflow flag and increment one or more registers to keep track of timer overflows.

The software delay loop that determines the counter gate time is slightly more complicated than the delay loop in the LED flasher project. The delay loop subroutine is a triple nested loop. Each time the innermost loop is executed, a subroutine is called that checks the timer overflow flag. If the timer has overflowed, the flag is immediately reset and one of the PIC's general purpose file registers is incremented to keep count of the overflows. If the overflow-count register overflows, yet another PIC file register is incremented. This increases the count to 32 bits (2E32), which allows a maximum count of more than four billion! 1Hz resolution up to 4GHz or 1kHz resolution up to 4THz is probably more than you will ever need.

At the end of the gate period, the counter gate is closed by making RA3 an output. This will clamp the input to the prescaler and freeze the count at its present value. The count will be stored as 32 bits in four of the PIC's file registers. The registers are defined as byte3, byte2, byte1 and byte0 in the program source code. The two most significant (highest) bytes of the count result are in byte2 and byte3. Any Timer0 overflows during the gate period will have incremented the value in these registers, so they already contain the correct values. Byte 1 of the count result is stored in the TMR0 register. As this register is directly readable, it is simple to extract this value and store it in byte1. The PIC instructions used are in **Box 3.4**.

That leaves us with the tricky problem of extracting the least significant byte of the count result from the prescaler. The prescaler is not directly readable. But Microchip Application Note AN592 [24] describes a clever way of extracting the prescaler value. We don't know the current value in the prescaler, but we do know that it is an 8-bit counter and it will overflow after it reaches a count of 255 decimal (FF hex). Each time the prescaler count reaches 255, it will increment the TMR0 register on the following pulse. We can recover the value from the

```
movf      TMR0,w     ;Get the value in TMR0
movwf     byte1      ;Store it
```

Box 3.4. The PIC instructions for extracting the value from TMRO register and storing it in byte1.

```
comf      temp,f      ;Yes, the prescaler has overflowed
incf      temp,w      ;Negate the pulse count and -
movwf     byte0       ;store it in byte0
```

Box 3.5. Code for complementing the pulse count and add one to the result.

```
movlw     0x1C        ;store the IF offset
movwf     count1
movlw     0x5A
movwf     count2
movlw     0x89
movwf     count3
movlw     0x00
movwf     count4
movf      count1,w
addwf     byte0,f     ;Add the first byte
movf      count2,w
btfsc     STATUS,C
incfsz    count2,w
addwf     byte1,f     ;Add the second byte
movf      count3,w
btfsc     STATUS,C
incfsz    count3,w
addwf     byte2,f     ;Add the third byte
movf      count4,w
btfsc     STATUS,C
incfsz    count4,w
addwf     byte3,f     ;Add the last byte
```

Box 3.6. Code for adding the commonly used IF frequency 9.001500MHz to the count before it is displayed.

prescaler by using I/O pin RA3 to provide a stream of pulses to the prescaler input. By counting the number of pulses required to overflow the prescaler, we can easily calculate the number that was originally in the prescaler. This is achieved by storing the original value of TMR0 in one of the file registers. A register that keeps count of the pulses is incremented and a pulse is sent to the prescaler input via RA3. The current value of TMR0 is then compared with the original value to see if it has changed. This process is repeated again and again until the prescaler overflows. Now that we have an accurate count of the number of pulses required to overflow the prescaler, we can calculate that the least significant byte of our frequency count is: (255-n)+1 where n = the pulse count. A fast and easy way to achieve the same result is to complement the pulse count (turn all binary ones to zeros and zeros to ones) and add 1 to the result. The code in **Box 3.5** was used to perform this function. Now that all four bytes of the count result are safely stored in the count registers, we can proceed to the next stage.

DISPLAYING THE FREQUENCY. Conventional frequency counters (Figure 3.5) have the count result 'hard wired' into the counter circuitry. If the counter is to be used as a frequency display for a superhet receiver or transmitter, it will be necessary to add or subtract an IF offset. Different offsets will be required for different operating modes. Some frequency mixing schemes will need to have the IF offset added to or subtracted from the local-oscillator frequency. Other systems will need to have the local-oscillator frequency added to or subtracted from the IF frequency. This can turn an already complicated circuit into a hideous nightmare of preset up/down counters and/or adders.

The PIC's ability to perform simple arithmetic means that it is very easy for us to compensate for IF offsets or to manipulate the count result in any way we want. The code in **Box 3.6** will add the commonly used IF frequency 9.001500MHz to the count before it is displayed. Note that 9001500 decimal is 895A1C hex. It really is that simple to add or subtract any IF offset you want. We won't be using this feature now, but it will be useful in future projects.

Before the frequency can be displayed on the LCD display module, it must be converted to a suitable format. The 32-bit binary number is first converted to packed BCD using a modified version of the binary-to-BCD library routine from Microchip's Math Utility Routines **[25]**. Each BCD digit is then converted to ASCII and sent one digit at a time to the LCD display module. Leading-zero suppression is implemented in the PIC program. This means that a frequency of 1MHz will be displayed as 1.000,000MHz and not 001.000,000MHz. The display is a two-line by 16-character intelligent LCD display module. I tested several different types of 2x16 display with the frequency counter. The Densitron display (Photo 3.5) is based on the Hitachi HD44780 LCD controller chip. This type of display uses an 8- or 4-bit parallel interface to the microcontroller. To avoid using up too many of the PIC's I/O pins, we will use the display in 4-bit mode. The module also has three control lines: ENA(enable), RS(register select) and R/W (read/write). The Maplin N27AZ display (**Photo 3.6**) is based on the Samsung KS0070B controller chip. This chip is compatible with the Hitachi controller. The Maplin display has a very effective LED backlight. The N27AZ data-sheet **[26]** has detailed information about the Maplin display module. The DC supply to the LED backlight should be current limited by a series resistor. If you take the DC supply for the backlight from the 7805 regulator, you will need to use a heatsink on the regulator. The Maplin display works well with about 100mA to the backlight LEDs.

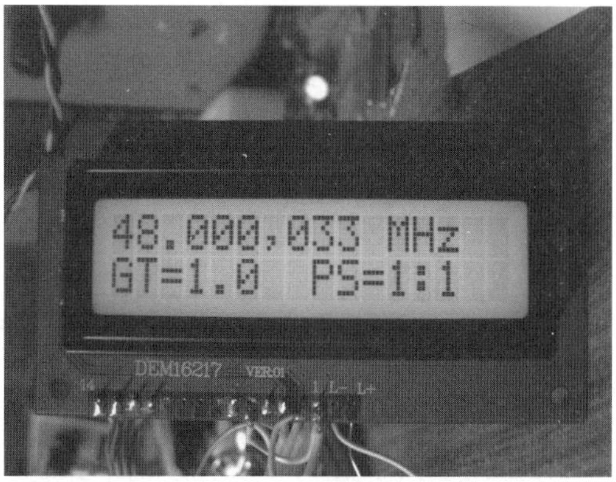

Photo 3.6: The Maplin N27AZ display showing 48MHz.

CONSTRUCTION. The circuit (**Figure 3.7**) is based on the schematic of the LED flasher. The counter can be built on an etched PCB or on Veroboard. The layout of my PCB is shown in **Figure 3.8**, the assembled circuit can be seen in **Photo 3.7**. The timebase oscillator is the most critical part of the circuit. Since the counter gate time is determined by a software delay loop, I could have used any old junk-box crystal, running at any frequency up to the PIC's maximum clock frequency of 20MHz. I decided not to use a standard CPU clock frequency like 4, 8, 12 or 16MHz because cheap CPU clock crystals have relatively poor frequency stability. I used a clock frequency of 10.240MHz because this is a standard PLL reference frequency. Crystals cut for this frequency will usually have lower frequency drift than CPU clock crystals. 10.240MHz crystals can be bought for less than 50p on internet auction sites. Most CBs use a 10.240MHz crystal in the PLL reference oscillator. A crystal that has seen several years service in a CB will be pre-aged and is quite likely to offer better stability than a brand new crystal.

I used 22pF ceramic capacitors for C5 and C6. You may find that you need to change these values to suit your crystal. The LCD display module is connected to the counter PCB by two lengths of ribbon cable. The cable for the data connections D4-D7 has four wires. The cable for the power, RS and ENA connections has five wires. Since we will only be writing to the display, the R/W connection on the LCD module is connected to ground by soldering a short length of wire between pin 1 and pin 5 on the display. Connection details are shown in **Figure 3.9**.

CALIBRATION. There are two way of calibrating the counter. You can use C7 to set the PIC's clock oscillator to exactly 10.240000MHz, or apply a signal of known frequency to the counter and adjust C7 until the correct frequency is displayed. Either way, you will need a reliable reference signal. If you have a receiver with an accurate frequency display, you can set

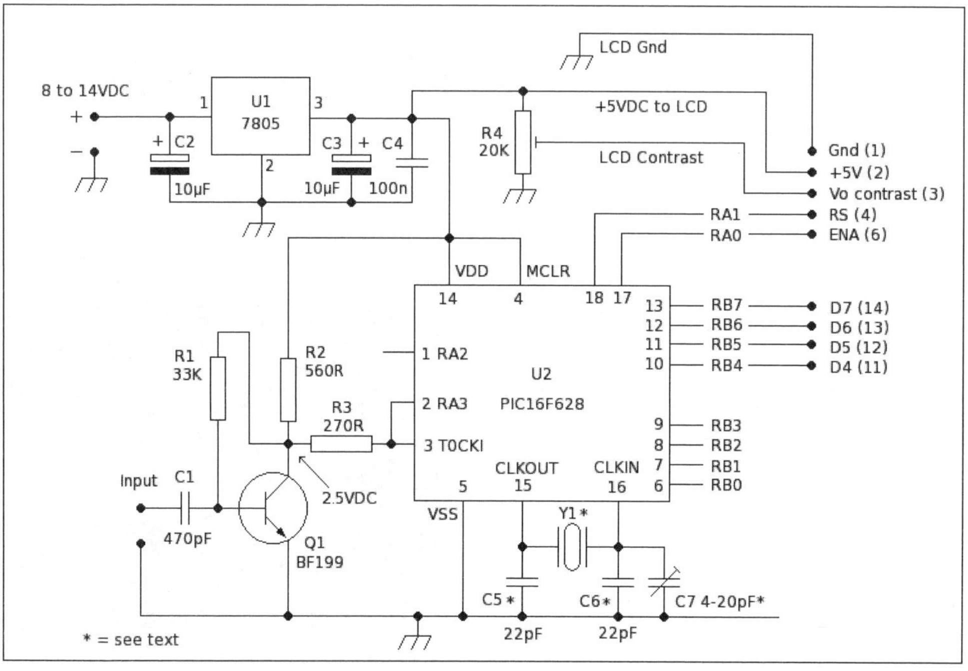

Figure 3.7: A schematic of the frequency counter.

it to 10.240MHz SSB and adjust C7 until the PIC's oscillator is zero-beat with the receiver. I use a 9.996MHz crystal oscillator that is carefully adjusted to zero-beat with the Russian atomic clock transmission RWM. The 9.996,000MHz signal from the oscillator is fed to the counter input and C7 is adjusted until the displayed frequency is correct. I used a 10.000MHz crystal with a series inductance to pull the oscillator frequency down to 9.996MHz.

A typical mid-range desktop or handheld frequency counter will have a claimed accuracy of about one part per million. More expensive counters will have a TCXO (temperature controlled crystal oscillator) or OCXO (oven controlled crystal oscillator) and frequency accuracy of one part in many millions.

TESTING. To test the accuracy and stability of the counter, it was connected to the output of my oven-controlled 10MHz frequency standard. After a warm-up of five minutes, the indicated frequency was 10.000,000MHz with an occasional count of 9.999,999MHz. I checked the displayed frequency every 30 minutes for the next 10 hours. There was no measurable drift during this period. The shack temperature was 24°C at the beginning of the test period. At 0730 the following morning, the shack temperature had fallen to 18°C and

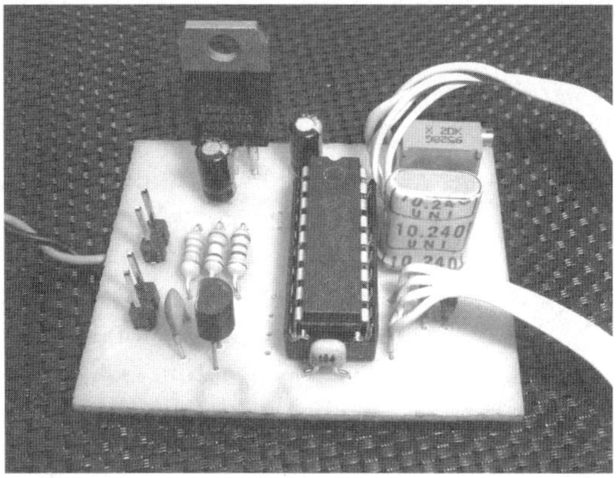

Photo 3.7: The assembled frequency counter.

Photo 3.8: The etched counter PCB.

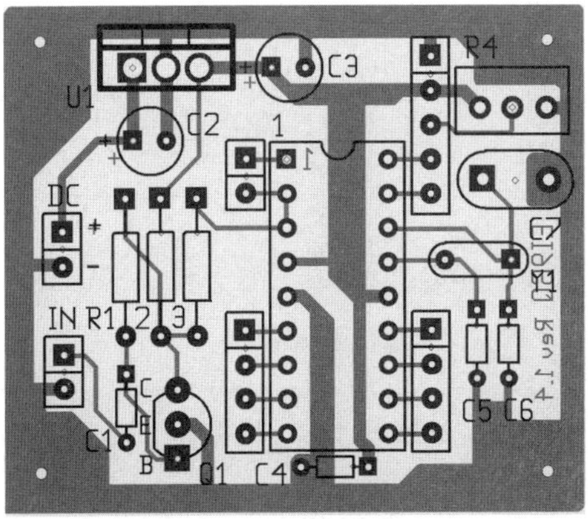

Figure 3.8: PCB layout for the frequency counter circuit.

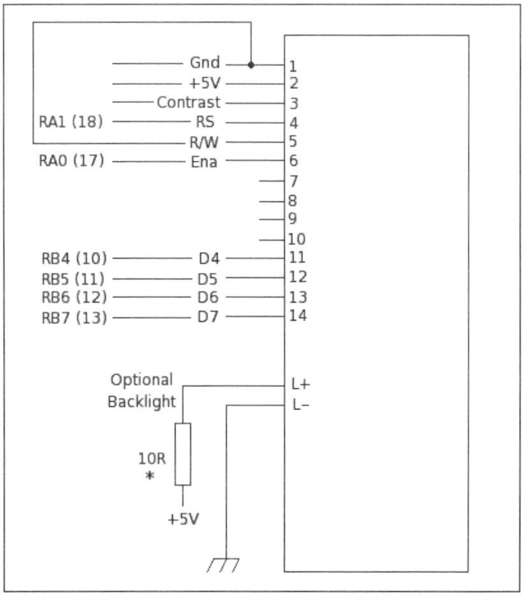

Figure 3.9: The connections to LCD module.

the indicated frequency was 9.999,997/8. By mid afternoon, the temperature was back in the mid-20s and the frequency count was back to exactly 10.000,000/999. The counter was then powered down for several days. When the counter was powered up and warmed up for five minutes, the indicated frequency was still 000/999. The greatest deviation from the correct frequency was -3Hz/+0Hz. It would not be reasonable to expect an accuracy of better than 0.3ppm from such a simple instrument. However, I have found that accuracy of better than 1ppm is easily achieved, provided that the counter is not subject to extreme temperatures or mechanical shock and the calibration is checked regularly.

FURTHER DEVELOPMENT. A bench frequency counter will always be an unfinished project. When choosing an enclosure for the counter, you should make sure you leave room for future additions. You will probably want to have two inputs, one for the normal HF input and the other for a VHF/UHF prescaler. You might want to replace the standard crystal oscillator with a high-stability OCXO or TCXO module. Note that several of the PIC's I/O pins are left unassigned. These I/O pins can be used to add extra functions to the counter. My Veroboard prototype of this counter has a very stable TCXO that was pulled from some scrap telecommunications equipment. This version of the counter has two inputs, one with a 1:4 prescaler (74F74) and switchable fast/slow (0.4/4.0 seconds) gate time and the other with a 1:128 UHF prescaler (MB505) and switchable 1.28 and 12.8 second gate time.

The PIC source code and PCB artwork are available for download from:

http://homepage.eircom.net/~ei9gq/counter.html

To avoid any confusion about copyright and licencing, I have released the counter program under the GNU 'General Public License'. Full details are included in the ZIP file.

FREQUENCY COUNTER COMPONENT LIST

U1	7805
U2	PIC16F628 (A or -20 version works best)
Q1	BF199
R1	33kΩ
R2	560Ω
R3	270Ω
R4	20kΩ (not critical) multi-turn pot.
C1	470pF
C2,C3	10µF
C4	100nF
C5,C6	22pF (see text).
C7	20pF 5mm trimmer cap. (2 pin type)
LCD module	Maplin N27AZ or similar 2x16 character display

Frequency counter improvements

When I built the original circuit for the frequency counter, my intention was to keep the circuit hardware as simple as possible and as close as possible to the trivial LED flasher.

The extreme simplicity of the circuit imposed a few limitations on the usefulness of the counter. The fixed gate time of 1 second is ideal for use in a bench frequency counter, but it is far too slow for use as a frequency display in a receiver or transceiver. A 100ms gate time version of the firmware can be found at: http://tinyurl.com/q24xj. A user selectable fast/slow gate time option would be a useful addition to the counter. Another limitation of the simple counter is a maximum count frequency of around 50MHz. This is acceptable for constructors of HF equipment, but not much use to a VHF/UHF enthusiast. The counter has a crude and simple input amplifier and gate circuit which consists of a single transistor amplifier and a resistor which is used as part of the gate circuit. My own preference is for a counter with a relatively low input impedance because I have found that this arrangement leads to less pickup of noise and mains hum. Many users would prefer a high impedance input which offers greater sensitivity and less loading of the circuit under test. The original circuit has no input protection. This is not a major problem for me because I don't often need to measure frequency at high power levels. Even if I was unfortunate enough to damage the input circuit, replacement transistors are cheap and easy to find. The new counter will have an optional high impedance input that can be used in place of the simple transistor circuit.

THE CIRCUIT. The schematic of the new front-end for the counter is shown in **Figure 3.10**. A UHF transistor is used as the input amplifier. A 74F74 dual D flip-flop is used as a 1:4 prescaler for the HF input. Most manufacturers specify a maximum clock input frequency somewhere between

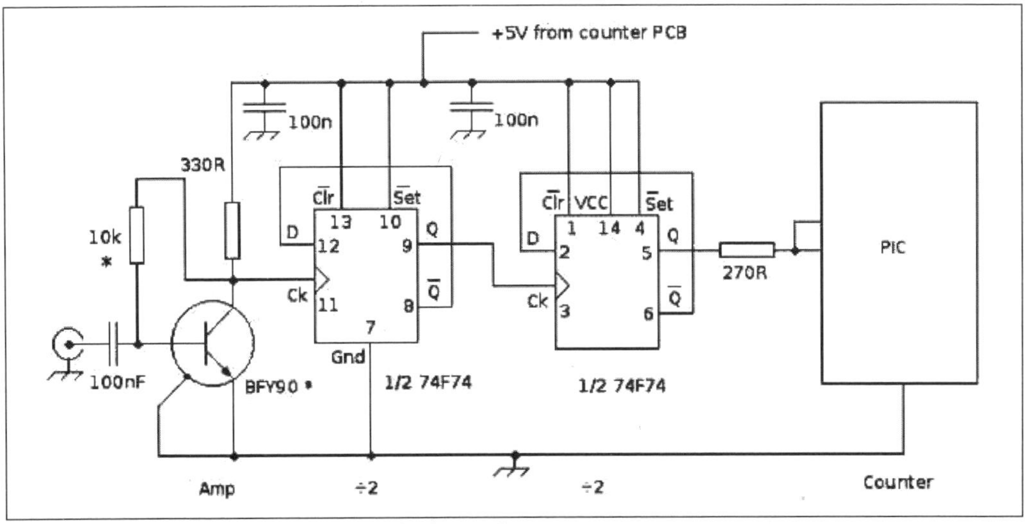

FIGURE 3.10: The counter's new front end

125MHz and 160MHz for the 74F74. I tested this circuit with several different Fairchild and Philips 74F74s and found that all of them would work up to at least 150MHz. Most of the main counter circuit is exactly the same as the original PIC counter.

One end of the 270Ω resistor (R3) of the counter PCB is unsoldered from the board so that the TTL level pulses from the prescaler can be fed to the PIC counter/timer input via this resistor. Transistor Q1 from the original circuit is no longer required and it can be removed from the board. **Figure 3.11** shows an alternative high impedance input stage with a FET source-follower as the first amplifier stage. The input is clamped to ±0.6 volts by a pair of 1N4148 diodes. The 39K resistor in series with the input limits the current through the protection diodes when the counter is driven from a low-frequency, high voltage source. This circuit will protect the FET from damage for input levels of several volts or even several tens of volts at the lower end of the HF spectrum. It is absolutely NOT suitable for use at mains voltage or any other source of dangerously high voltages. As this resistor is bypassed by a 47pF capacitor, it has little or no effect on the counter input sensitivity at the VHF end of the spectrum.

RESOLUTION. The use of a 1:4 prescaler at the counter input means that the PIC counter will show a frequency that is exactly one quarter of the input frequency. To correct this error, we must increase the gate time so that it is four times longer than it was in the original circuit. This results in a gate time of 4 seconds for a count resolution of 1Hz. This is a bit too long, even for a high precision bench counter. The new counter will offer selectable gate times of 0.4 seconds for a count resolution of 10Hz and 4 seconds for 1Hz resolution. The longer gate time allows the counter to have a count resolution of 1Hz, even on 6m, 4m and 2m. This is a very useful feature for the VHF experimenter. Because the counter gate time is defined by software, it is quite easy to add more gate time options if they are needed. The counter will have a separate input and prescaler for UHF and lower microwave frequencies.

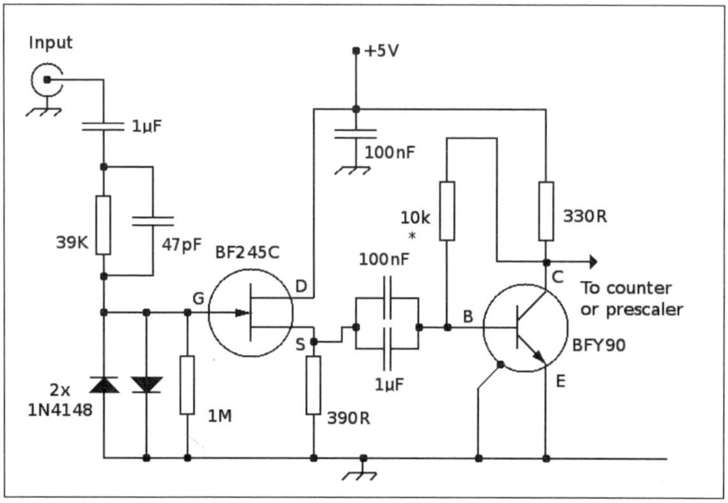

FIGURE 3.11: The alternative high impedance input stage.

CONSTRUCTION. The counter front end consists of the input amplifier and 1:4 prescaler which was built dead-bug style on a small sheet of PCB laminate. **Photo 3.9** shows the method of construction. **Photo 3.10** shows the completed prototype. The high impedance input circuit is on the left, and on the right of the photo is the simple PIC counter PCB. The FET is a BF245C (Maplin N65AG). Any similar VHF N-channel J-FET should work equally well. The UHF bipolar transistor is a BFY90. This device has a lead which is connected to the metal case of the transistor. This lead should be connected directly to ground. A similar UHF NPN transistor can be substituted for this part. Check that the voltage at the collector of this transistor is approximately 2 to 2.5VDC when the circuit is powered up. The 100nF capacitors are ceramic types. The 1μF capacitors are plastic types. The voltage ratings are not too critical. The capacitors at the counter input should be able to withstand the maximum voltage that the counter is likely to see at the input. Capacitors with a 50 to 100V rating should be fine. All resistors are standard 0.25W carbon film types.

PHOTO 3.9: Construction of the counter's front end.

PHOTO 3.10: The prototype counter.

A 10.240MHz crystal is used for the PIC clock oscillator. This is a good choice for use as a frequency counter timebase because just about all 10.240MHz crystals are designed for use in a PLL reference oscillator. 4, 8, 10 and 12MHz CPU clock crystals are cheaper and easier to find, but they are unlikely to be as stable as a typical 10.240MHz unit. I am planning to modify the firmware so that an external 10.000MHz oscillator can be used instead. More about this later.

The 5V regulator on the counter PCB needs a heatsink because it has to provide current for the LED backlight in the LCD display module. Modified PIC firmware for the frequency counter is available on the web http://tinyurl.com/q24xj.

TESTING. The only adjustment required is the alignment of the PIC clock oscillator. The procedure is described in the original frequency counter article. I used a 9.996MHz crystal oscillator which was carefully tuned to zero beat with the RWM atomic clock signal from Russia. This station transmits a steady carrier for 8 minutes starting exactly on the hour and again on the half hour. The signal is very strong with very little QRM or fading from sunrise to well after midday during the winter months. It is trivially easy to adjust the test oscillator to within 1Hz of the RWM carrier. This is an error of one part in 10 million. With a little care and patience, even greater accuracy is possible. The counter accuracy was also checked against my 10MHz oven controlled frequency standard which is maintained to an accuracy of better than $2x10^{-8}$. There is little point in trying to calibrate a counter with a simple crystal oscillator timebase for accuracy greater than one part in a few million. I checked the accuracy of the counter over a period of several days. The original counter was remarkably stable.

PHOTO 3.11: The counter display shows frequency, prescaler division ratio and counter gate time.

Although this unit is identical to the original, I found that the long term stability of the new counter is not quite as good. However, the stability is still reasonably good at better than 1ppm. The greatest error measured was on a very cold morning when the indicated frequency was out by 6Hz at 10MHz. The slight difference in performance between the two counters is probably due to differences in quality and temperature coefficient between the two 10.24MHz crystals and the crystal oscillator shunt capacitors on the counter PCB.

The counter was also tested at VHF frequencies up to 146MHz. I found that the counter was able to 'sniff' RF from a nearby 2m transmitter running at 145.995MHz. The VHF count was always accurate and reliable. I used a sine wave function generator to test the low frequency performance of the counter and found that the counter was reliable down to below 100kHz. I was hoping that the prescaler and counter would work down to audio frequencies, but it seems that the high-speed 74F74 flip-flop does not like the slow rise and fall times of very low frequency signals. Ah well, I suppose 100kHz to 150MHz is not bad for a HF counter. **Photo 3.11** shows the display when measuring the frequency of a 116.553MHz oscillator.

MODIFICATIONS AND EXPERIMENTS. Overall, I was quite pleased with the performance of the frequency counter. The only thing that that annoyed me slightly was a tendency for the last digit of the counter sometimes to read '1' instead of '0' when there was no input signal. After some head scratching, I figured out that the problem was. When a signal is applied to the counter input, the output of the 74F74 prescaler toggles up and down at a rate of one complete cycle for every four cycles of input signal. If the input signal is removed, there is a 50% chance that the prescaler output will stop at a high output level. Each time the PIC counter gate is opened, the high output on pin 5 of the 74F74 creates a single positive pulse which increments the count from zero to one. This problem was easy to solve by using one of the reset (Clr) inputs of the 74F74 to ensure that the prescaler output is always low at the beginning of each count. Since the reset input will always pull the D-flip-flop output low regardless of the state of the Clk and D inputs, the reset input can also be used as the counter gate. This replaces the crude gate mechanism used in the original PIC counter. A single wire from pin 1 of the PIC (RA2) to pin 1 of the 74F74 provides the gate/clear signal. It only took about five minutes to add a few lines of code to the PIC program to implement the new gate function. Note that the Clr input of the 74F74 is active low. This means that pin 1 of the PIC is high during the count and the counter gate is closed when this pin goes low. The modified counter always shows a frequency of zero in the absence on an input signal.

I have also built a UHF prescaler based on a divide-by-128 ECL divider. The prescaler is based on the Fujitsu MB505 IC which is rated for a maximum input frequency of 1.6GHz. I have tested this circuit at various frequencies up to 1300MHz and found that it is well behaved and reliable.

UHF Prescaler for the Frequency Counter Project

FREQUENCY MEASUREMENT. Frequency is a measure of the number of events that occur within a given period of time. For our purposes, the events to be measured are single cycles of alternating current or voltage, and the unit of time is usually one second.

The input signal to the frequency counter has been carefully conditioned so that it presents a logic level square wave signal to the counter circuit. The train of square wave pulses is easily and accurately measured using a digital counter. Establishing the exact timing of the gate control signal is a rather more difficult problem. Before we get to the matter of gate timing, we should first consider the exact meaning of the word time. Two of the many definitions of time offered by dictionary.com:

"– noun
1. the system of those sequential relations that any event has to any other, as past, present, or future; indefinite and continuous duration regarded as that in which events succeed one another.
...
4.a limited period or interval, as between two successive events: a long time."

Historically, time was measured relative to the rate of rotation of the Earth. Night and day were subdivided into smaller units like hours minutes and seconds. In more recent times, time has been measured relative to the period of the Earth's orbit around the Sun. Since the 1960s, the SI (International System of Units) unit of one second has been very precisely defined as: "*the duration of 9,192,631,770 periods of the radiation corresponding to the transition between the two hyperfine levels of the ground state of the caesium-133 atom"* **[27]**. Atomic clocks that keep time based on this natural resonance in the Caesium atom are known as primary frequency standards. Such clocks are accurate to within one second in several million years. An international network of atomic clocks is used to maintain UTC (Coordinated Universal Time) and several other timescales. Many of these clocks are used to give very precise control of the frequency of radio transmitters so that they can be used as frequency standards. RWM (Russia), WWV (USA), MSF (UK) and other frequency standard transmissions have a carrier frequency which is maintained to a very high level of accuracy. For example, the 60kHz carrier of the MSF time and frequency signals from Anthorn is maintained to an accuracy of less than 2 parts in 10^{12}. Most time and frequency stations use phase and/or amplitude modulation to add additional information to the transmitted signal. These variations must be averaged out over a relatively long period of time if the carrier is to be used as a reference for a local frequency standard.

It is not possible to establish the exact transmission time or carrier frequency of a remote radio transmission because of the propagation delay of the signal. Radio signals propagate at almost 300,000km/s or just under 300 metres per microsecond. If you know the precise distance between your receiver and the transmitter of the remote frequency standard, it is possible to make a rough estimate of the propagation delay. Other variables like phase shifts caused by reception of multi-path signals and Doppler shift caused by changes in the ionosphere are almost impossible to predict. The GPS satellite navigation system provides a ready source of time and frequency information. Each of the satellites in the GPS constellation has a very accurate atomic clock which is used to control the frequency and timing of the signals transmitted by the satellite. The short term accuracy of the timing pulses from a typical GPS receiver is not particularly good, but if the errors are averaged out over a long period of hours or even days, it is possible to build a GPS controlled frequency standard with accuracy approaching that of an atomic clock.

SIMPLE FREQUENCY MEASUREMENT TECHNIQUES. The standard way of calibrating a frequency standard oscillator is to compare its frequency with a trusted reference. By comparing the phase of the oscillator's output signal with the phase of the reference signal, it is possible to establish the frequency difference between the two. If both oscillators are running at exactly the same frequency, the phase difference between the two signals will remain constant. Any frequency difference between the two oscillators will result in a change in the relative phase difference between the two signals. The reference signal usually comes from a remote time and frequency transmission. We can use a standard radio receiver to receive both signals. Ideally, the strength of the locally generated signal should be attenuated until it is identical to the strength of the remote signal. This is not always possible in practice because the signal from the remote station is subject to fading.

When the local signal is in phase with the signal from the remote station, the amplitude of the received signal is at a maximum. When the local signal is 180° out of phase with the remote signal, a null can be heard in the received signal. By timing the interval between these swells and nulls, it is possible to establish the frequency difference between the two signals. If you happen to live within ground wave distance of a frequency standard station, this method can be extremely accurate. If the frequency standard station is more than a few miles away, the signal will be subject to short term variations in phase and frequency. If the local oscillator is stable enough, these variations can be averaged out over a period of time. In practice, I have found that I can easily adjust my local frequency standard oscillator to within 1Hz of RWM within a few seconds and within 0.2Hz (one null every five seconds) over a period of several minutes. This corresponds to an accuracy of one part in 50 million. My 9.996MHz oscillator is only stable to within 1Hz per hour after a warm up period of 24 hours. This means that my reference can only be trusted for a few minutes after the RWM carrier disappears. This allows just enough time for it to be used to check the calibration of my main oven controlled reference oscillator.

The human ear makes a very efficient phase detector and the human brain is very good at filtering out sudden random phase changes caused by changes in the ionosphere from the periodic phase changes that are caused by frequency differences between the local and remote oscillators. However, there are even more accurate methods of comparing the phase of two signals. If the two signals are monitored on a dual beam oscilloscope, the phase difference between the two signals can be monitored in real time without having to wait for the 0 and 180° swells and nulls of the zero beat method. **Photo 3.12** shows an oscilloscope display of the signal from two oscillators. The horizontal timebase of the scope is triggered by the signal shown in the upper trace. This means that the upper trace remains static on the oscilloscope display. The source of the upper trace signal is a 10MHz OCXO. The lower trace is a 5MHz signal from a 5MHz OCXO. The lower trace is not synchronised with the X timebase so that it is free to wander in and out of phase with the upper signal. After a warm up period of several hours, the time taken for the bottom trace to drift by 180° ranges between 20 seconds and one minute. This is a drift rate of 1 part in several hundred million. **Photo 3.13** shows some of the Lissajous patterns that are seen when the two OCXO signals are fed to the X and Y inputs of the scope.

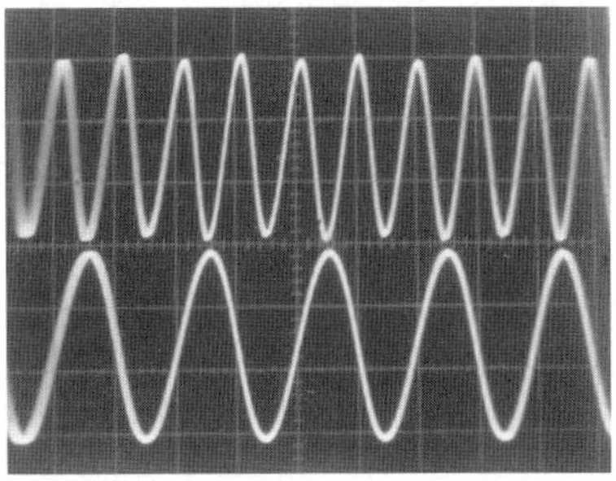

PHOTO 3.12: Using a dual beam oscilloscope to compare the phase of two oscillators. The signal on the upper trace is twice the frequency of the lower one.

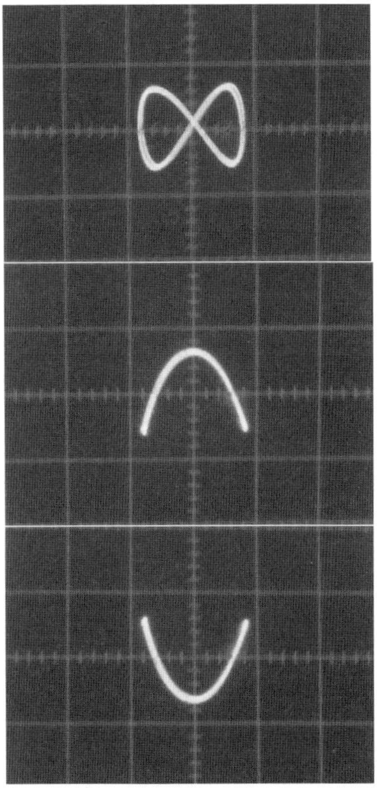

PHOTO 3.13: Typical Lissajous figures.

Radio amateurs have used many different signal sources as a reference for frequency standards. There are several published designs for off-air frequency standards which are phase locked to the MSF signal on 60kHz or the BBC long wave transmitter on 198kHz. See the excellent articles by Andy Talbot, G4JNT **[28]** and James Miler, G3RUH **[29]** for further information about PLL frequency locking methods. There are also some published designs which use the line sync pulses from a television receiver as an accurate frequency reference. This method is only of historical interest because of the planned closure of the analogue TV transmission network – the signals from set-top boxes are nowhere near as accurate.

EXTENDING THE FREQUENCY RANGE OF OUR COUNTER PROJECT. The easiest way to extend the frequency range of any frequency counter is to use a prescaler (divider) in front of its input. If you had a counter which only extended to 10MHz you could put a suitable a divide-by-10 prescaler in front of it which, if fed with 100MHz, would output just 10MHz, thus extending the useful range of your counter. We are going to apply a similar technique to extend the range of our counter project.

High speed emitter-coupled logic (ECL) prescalers are widely used in both terrestrial and satellite TV receivers as part of the tuning system. They can also be found in the phase locked loop (PLL) synthesisers used in mobile phone and other types of UHF and microwave telecommunications equipment. Popular devices like the Motorola 12xxx and Fujitsu 5xx series are no longer in production, but they are still readily available from specialist suppliers **[30]** and internet auction sites. The prescaler used in this month's project is a Fujitsu MB505. This device is available in an 8 pin DIP or surface mount package. Either type of IC package is well suited to dead-bug style construction. Other Fujitsu 5xx devices like the MB506 (2.4GHz) and MB510 (2.7GHz) can be used with very little modification to the circuit. Alternatively, you can use one of the many different 1:128 prescaler ICs from other manufacturers like Motorola, NEC or Philips. Putting a divide-by-128 prescaler on the front of a counter

The MB505 is rated for a maximum input frequency of 1.6GHz and a minimum input frequency of 100MHz. In practice, I have found that the circuit counts reliably from 5-10MHz up to 1296MHz. As I don't have a suitable signal source, I was unable to test the prescaler's performance at 1.6GHz and beyond. **Figure 3.12** shows a simplified diagram of the MB505 internal circuitry. The UHF input is applied to the positive edge triggered input (pin 1). There is also a complementary negative edge triggered input (pin 8) which is not used in this application. The counter/divider is an 8 bit binary ripple counter of the type described previously. One of the eight flip-flops is switched by an external control voltage applied to pin 3, this allows the circuit designer to select a divide ratio of 1:128 or 1:256. The prescaler output is from pin 4. Note that the output requires an external load resistor. ECL logic outputs usually have a much lower voltage swing than their TTL or CMOS counterparts. The output from the MB505 is typically about 1V p-p. This is below the level required for reliable operation of standard TTL or CMOS logic.

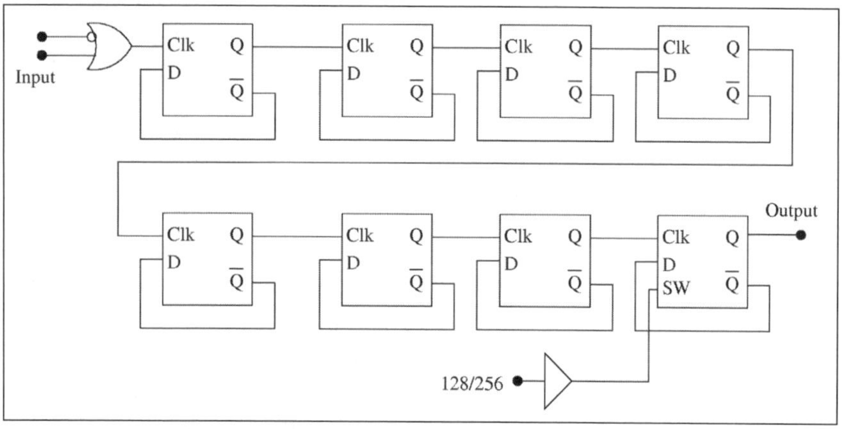

FIGURE 3.12: Simplified diagram of the MB505.

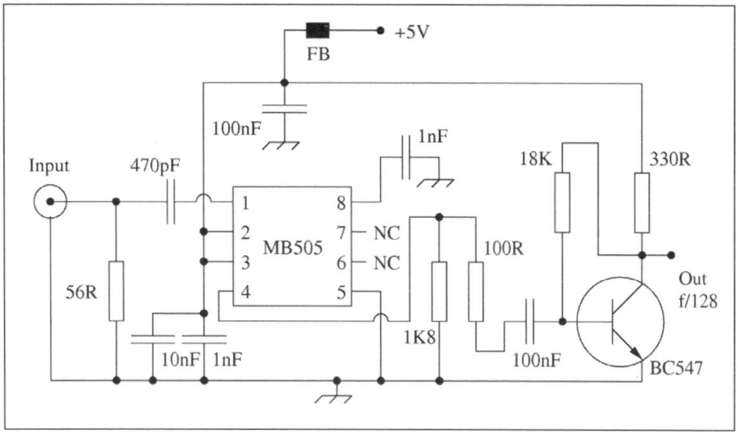

FIGURE 3.13: The practical prescaler circuit.

THE CIRCUIT. The schematic of the prescaler is shown in **Figure 3.13**. The UHF part of the circuit is almost identical to the test circuit used in the Fujitsu prescaler datasheets **[31]**. The SW input is tied to the 5V Vcc supply so that the division ratio is 1:128. The output from the prescaler is amplified by a transistor amplifier which produces a nice clean square wave output of almost 5V p-p.

CONSTRUCTION. The circuit was built dead-bug style on copper PCB laminate. **Photo 3.14** shows the layout of the circuit. It is critically important to keep all ground connections as short as possible. In the GHz region, even a couple of mm can make a real difference to the stability of the circuit. Ordinary ceramic disc capacitors can be used provided that the leads are trimmed to a length of no more than 2-3mm. Preferably, the 1nF decoupling capacitors should be leadless surface mount chip capacitors.

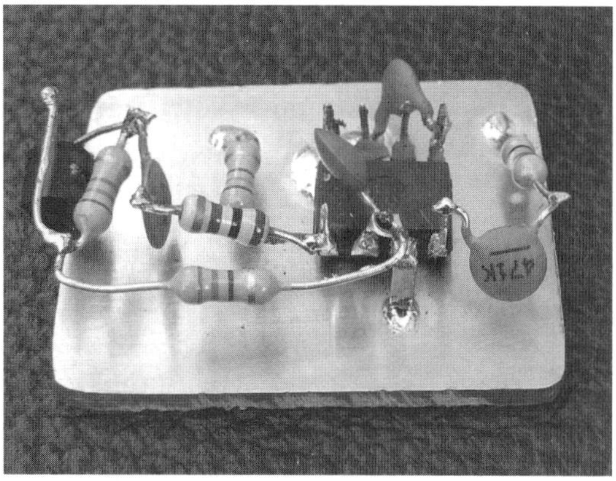

PHOTO 3.14: Dead-bug version of the prescaler circuit.

The prescaler unit should be mounted behind the BNC socket on the counter front panel. If you use a square type of BNC socket with four fixing holes, you can use a pair of solder tags on two of the screws to provide a convenient mounting point for the prescaler board. This approach ensures a short, low inductance path to ground and a very short connection from the centre pin of the BNC socket to the prescaler input. The main counter unit and the HF prescaler were described last month. If you don't want to go to the trouble of etching a PCB for the main counter board, you can build the circuit on a strip of Veroboard. The layout of the main counter unit is not particularly critical. **Photo 3.15** shows a Veroboard version of the

PHOTO 3.15: The modified counter circuit built on Veroboard.

PHOTO 3.16: The finished counter.

counter. The tin can on the right is a very high quality TCXO (temperature compensated crystal oscillator) which was pulled from a surplus mobile phone base unit. This is used instead of the 10.24MHz crystal. A pair of DPDT switches (SW1, SW2) have been added to the circuit. One pole of SW1 is used to switch between the HF and UHF prescalers, the other pole of SW1 is used to 'tell' the PIC which input is in use. SW2 is used to select fast or slow gate times. Note that even though there are only two positions on the gate time switch, there are actually four gate times. When the HF input is selected, the gate times are 0.4s and 4.0s, the gate times for UHF are 1.28s and 12.8s. Count resolution is 10Hz and 1Hz for the HF input and 100Hz and 10Hz for the UHF input. I used an old PC parallel port switch box as an enclosure for the project. **Photo 3.16** shows the finished counter, while **Figure 3.14** shows the revised PIC counter schematic.

FIRMWARE. The latest version of the PIC software is available at http://tinyurl.com/q24xj. I have also written a modified version that is designed to work with a 10MHz clock oscillator. This will allow the constructor to have the option of using an external high stability oscillator instead of the PIC's built in oscillator. Surplus 10MHz OCXO (oven controlled crystal oscillator) modules are often available on eBay. A 5.000MHz OCXO and frequency doubler can also be used. For the ultimate in accuracy and stability, you could use a rubidium standard or a GPS disciplined oscillator as the PIC clock oscillator.

The modifications to the firmware were quite simple. I added a few lines of code to check the state of SW1 and SW2. Two new gate delay routines were written, one for 100ms which is called four times for the 400ms gate delay and one for 1.28s. The 1.28s delay is repeated ten times for the 12.8s gate. The original one second delay routine is called four times for the 4.0s gate. The gate and input selection code is shown below.

```
btfss   PORTB,inp_sel   ;Check PIC port pin for SW1
goto    UHF             ;Input B selected. Go to UHF routine.
btfss   PORTB,gatesel   ;Check PIC port pin for SW2
goto    HFfast          ;Use short gate time.
                        ;Otherwise do nothing. Use default - HF input and 4.0s gate
```

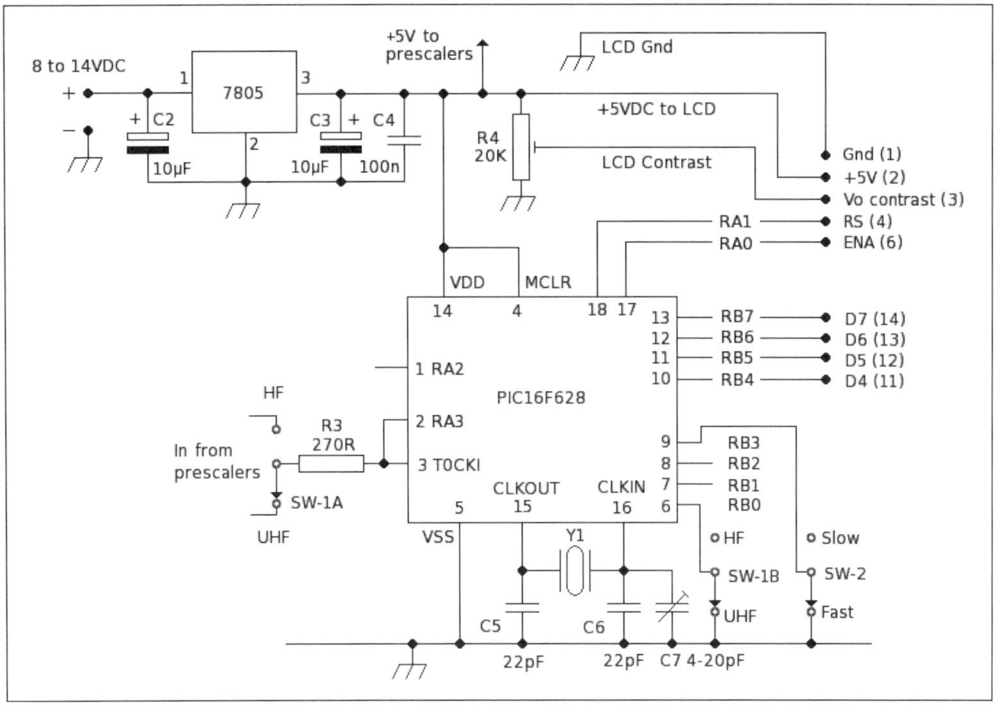

FIGURE 3.14: Revised PIC counter schematic.

TESTING AND CALIBRATION. As usual, I used a simple 9.996MHz crystal oscillator which was tuned to zero beat with RWM (A Russian broadcast frequency standard) to calibrate the counter. Calibration was also checked against my OCXO frequency standard. This was my third attempt at building a counter with a UHF prescaler. My earlier efforts were also based on Fujitsu IC's. Both of these prescalers have performed flawlessly for several years. Since I have enjoyed such great success in the past, I was quite confident that the new circuit would perform just as well. When the circuit was first powered up, I found that the HF input worked as expected, but when the UHF input was selected, the display indicated a constantly changing frequency in the 20-50MHz region, even when no signal was applied to the prescaler input! Clearly the prescaler was not stable under all conditions. However, I found that the counter worked correctly when an input signal of more than a few mV was applied to the prescaler input. This is a known problem with the 5xx series and other similar prescalers. When these devices are used for their normal application as frequency dividers in PLL synthesisers, this problem does not arise because there is always a signal present at the input of the prescaler. I rebuilt the circuit several times using both leadless chip capacitors and ceramic disc capacitors. This resulted in no improvement. I even dismantled my perfectly good bench counter which also has an MB505 prescaler to see if I could find the source of the problem, all to no avail. It seems the problem is caused by noise in the input amplifier of the IC. I eventually discovered that a very small DC bias on the input would completely cure the problem. Putting a 100kΩ resistor from pin 1 to ground stopped the instability. I increased the resistance to 470kΩ, 1MΩ and finally 2MΩ. The instability problem returned when the resistance was increased beyond 2MΩ. I finally settled on a value of 1MΩ. The prescaler is now unconditionally stable and the counter always shows 0.000.00MHz in the absence of an input signal. If you encounter the same instability problem, try putting a 1MΩ resistor from pin 1 of the prescaler directly to ground. The tiny bias current of about 2.5μA seems to have little or no effect on the normal operation of the prescaler.

ACCURACY AND STABILITY. The long-term accuracy of the counter will depend on the stability of the PICs clock oscillator. The simple crystal oscillator is accurate to well within 1ppm over a typical range of shack temperatures at EI9GQ. The counter is unlikely to achieve this level of accuracy if it is subject to more extreme temperature changes or mechanical shock. Very accurate counters usually use a very stable TCXO (temperature controlled crystal oscillator) or OCXO as a timebase oscillator. Since the PIC can use an external clock source, it is quite possible to use a high stability oscillator as a clock for our homebrew counter.

Anyone who has ever built a free-running LC VFO will know that changes in temperature are the main cause of frequency drift in oscillators. Even if your VFO is well constructed and mechanically sound, changes in temperature will lead to small changes in the values of key components in the VFO which will result in a change of output frequency. A crystal oscillator is usually much more stable than a VFO, but there will always be some tendency for the oscillator to drift. This drift cannot be totally eliminated so our goal is to keep it within acceptable limits. Our UHF counter with a maximum frequency of more than 1GHz and count resolution of 10Hz for signals applied to the UHF input would ideally be accurate to better than 1 part in 100 million. This level of accuracy and stability is well beyond the capabilities of a normal crystal oscillator. There are two basic methods of stabilising the frequency of a crystal oscillator. If we accept the fact that the oscillator will drift with changing temperature, we can try to correct this drift by using some form of temperature compensation. The other approach is to control the environment of the crystal oscillator so that it is protected from changes in ambient temperature. One way of achieving this is to place the crystal or perhaps even the entire oscillator and its buffer amplifier in a temperature controlled oven. The oven is operated at a relatively high temperature which is carefully controlled to within a fraction of a degree. Some of the best TCXOs are stable to well within 1ppm over the entire operating temperature range of the device. This is typically from 0°C to 50°C. Over a more typical range of room temperatures, say 10°C to 28°C, you could expect to achieve about 0.1ppm. High spec OCXOs are stable to about +/-1ppb (yes I really do mean parts per billion).

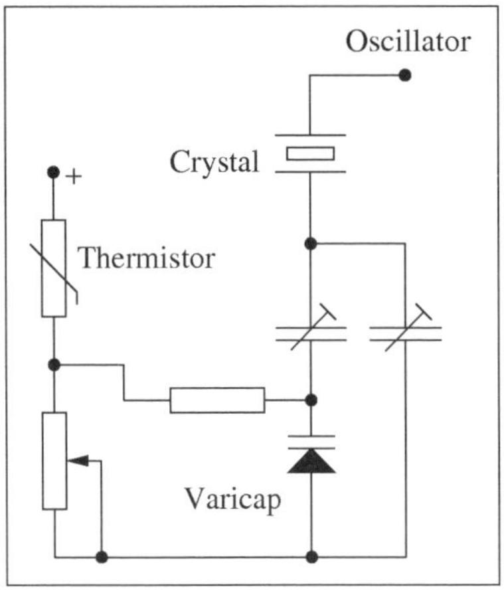

FIGURE 3.15: An idealised temperature compensation scheme that, unfortunately, won't work in practice.

Figure 3.15 shows a simple temperature compensation circuit which uses a thermistor to measure the temperature of the oscillator and a varicap diode to correct the frequency of the crystal oscillator. At first glance, this approach would appear to offer a cure for all thermal drift in crystal oscillators. It would seem that all we need to know is the precise amount of oscillator drift that is caused by a given value of temperature change so that we can choose a thermistor which has the correct temperature/resistance curve to fully compensate for the drift. This solution works up to a point, but it is not perfect. Thermistors don't have a linear resistance/temperature characteristic. Similarly, varicap diodes don't show a linear capacitance/voltage characteristic.

It is possible to use a very accurate temperature sensor in combination with a microcontroller, a DAC (digital to analogue converter) and a varicap diode to compensate for thermal drift. The cost of the microcontroller and DAC can be justified on the grounds that the alternative crystal oven is even more expensive and power-hungry. Even the most sophisticated compensation schemes are not perfect. Quartz crystals have many imperfections which lead to spurious resonances within the crystal, which tend to shift with age and temperature. The use of high speed digital circuits in close proximity to the analogue crystal oscillator can also introduce unwanted noise to the oscillator output. Digital control circuits must be very carefully isolated from sensitive analogue circuits like oscillators and amplifiers. Despite these limitations, TCXOs can offer a significant improvement in stability compared to simple crystal oscillators. Good quality TCXOs offer frequency stability which is at least an order of magnitude better than an uncompensated oscillator. State-of-the-art TCXOs can be almost as stable as a typical OCXO.

GPS Satellite Network to Control a Local Frequency Standard

High quality oven controlled crystal oscillators (OCXO) are readily available on the surplus market. This type of oscillator can be used to make a very useful frequency standard. OCXOs made by Isotherm, Vectron, C-Mac, Temex, HP and other manufacturers can offer part per billion frequency stability at a very reasonable price. My main shack frequency standard is a Temex 10MHz OCXO that was purchased on eBay. The cost was about £20 including P&P. The manufacturers data-sheet for this OCXO specifies a maximum frequency drift due to crystal ageing of 0.5ppb ($5x10^{-10}$) per day and frequency stability better than $1.6x10^{-9}$ over a temperature range of 0 to 50°C. These are worst case figures for a brand new unit. By the time these oscillators reach the surplus market, they have probably seen a few years of ageing and are likely to offer even better performance than a new unit. The date code on my Temex unit shows that it was manufactured in 2001 and a label attached by the original purchaser shows that it was burn-in tested in 2003. Measurements over the last few months have shown that drift due to crystal ageing is actually less than 1ppb per week. I also have an older 5MHz OCXO that was pulled from a 20th Century vintage marine satellite navigation system. The short term stability of this unit is not quite as good as the Temex unit, but the long term stability is excellent at $2x10^{-8}$ over a period of four years!

FREQUENCIES. OCXOs are often custom made for a specific application. An oscillator that was designed for use in a radio or TV transmitter might use a non-standard crystal frequency. Such oscillators can still be useful to the radio amateur. For example, you could use an oscillator with an odd frequency like 60.526MHz for calibrating a frequency counter or as a clock for a DDS (direct digital synthesiser) which could be programmed to produce a more useful output frequency. There are a few commonly used standard frequencies for OCXOs. Many OCXOs have an output frequency of 10MHz or 5MHz. Most commercially made off-air frequency standards have an output frequency of 10MHz.

Internet auction sites like eBay are by far the easiest way of finding a suitable oscillator. Prices for a typical 5 or 10MHz unit range from £10 to £30. Very high specification double oven (oven within an oven) types often change hands for more than £50. If an eBay search for 'ocxo' fails to locate a suitable oscillator, try searching for 'frequency standard' or 'crystal oven'. Rubidium (Rb) and Caesium (Cs) atomic frequency standards are often seen on eBay, but the asking price is usually well beyond the EI9GQ homebrew budget. Lower cost Rb units are usually ex-equipment and sold on a strictly 'sold-as-is' basis. Guaranteed working Rb standards are available at a cost of several hundred pounds each.

Once you have acquired a suitable OCXO, all that you need to build a frequency standard is a well screened enclosure, a power supply (usually 12V or 5V DC) and a simple buffer amplifier. Frequency dividers and/or multipliers can be used to provide outputs at multiples or sub-multiples of the OCXO frequency. My shack standard uses a 10MHz OCXO and a couple of divider ICs to produce outputs at 10MHz, 5MHz and 1MHz.

CALIBRATION. A very stable frequency standard is not very useful unless its output frequency is accurately calibrated. A very stable oscillator will remain on the same frequency, plus or minus a very small margin, for a long period of time. It is important to note that stability is not the same as accuracy. If our very stable oscillator is not adjusted to the correct frequency, its low frequency drift will only ensure that it remains on the wrong frequency for a long period of time! Before it can be used as a frequency standard, our oscillator must be carefully checked against a trusted standard.

Previously we discussed some methods of calibrating the oscillator in the frequency counter project using the RWM (Russian atomic clock) transmissions on 9.996MHz. These same methods can be used to calibrate a 10MHz standard. A 10MHz receiver can be used to zero-beat the signal from the 10MHz OCXO (or the second harmonic of a 5MHz OCXO) with one of the atomic clock signals. Unfortunately, this is very difficult to do because there is no 10MHz frequency standard transmitter here in Ireland. I can hear several clock transmissions on 10MHz at various times of the day. BPM from China usually has the strongest signal. I can also hear WWV from the USA and several other unidentified 10MHz carriers. Unlike the strong and steady RWM signal from Russia, the 10MHz signals are usually weak and noisy. The fact that there are several out-of-phase 10MHz signals combined with interference from adjacent stations and severe fading makes it almost impossible to get closer than about ±1Hz to the BPM carrier. This is an accuracy of $1x10^{-7}$ which is reasonably good for many purposes, but certainly not good enough to calibrate an OCXO standard.

To calibrate our frequency standard, we will have to find a more accurate source of time and frequency information. There are a few options available. You could use the carrier frequency of a local radio or TV broadcast station. This approach will only work if you can be absolutely sure that the broadcast station carrier frequency is maintained to a very high level of accuracy. Many off-air frequency standards are phase locked to the LF time signal on 60kHz or the 198kHz carrier of the BBC4 long wave transmitter. The 100kHz Loran navigation signals can also be used to control a frequency standard. Since GPS receivers are now very cheap and readily available, most of the newer time and frequency standard designs use the GPS satellite network as an accurate time and frequency reference.

GLOBAL POSITIONING SYSTEM. The NAVSTAR GPS satellite constellation provides a ready source of accurate timing information. Each satellite has a very accurate atomic clock. The accuracy of the on-board clocks is checked against atomic clocks at the GPS control ground stations to ensure the health and accuracy of the system. Each satellite transmits data which contains information about the precise position of the satellite and the time that the data was transmitted. This information can be used to calculate the distance between the satellite and the GPS receiver. By analysing the data from several satellites, the receiver can calculate time to within a few ns and the position of the receiver to within a few metres.

This ability of the GPS receiver to synchronise its own internal clock with the very accurate clocks on the GPS satellites makes it very useful as a time and frequency reference. There are several sources of error which will limit the accuracy of a GPS based clock. The clocks on the satellites will have some small frequency error. This error is partially offset because the data stream transmitted by the satellites contains information about the accuracy and health of the system. This information is based on observations at ground

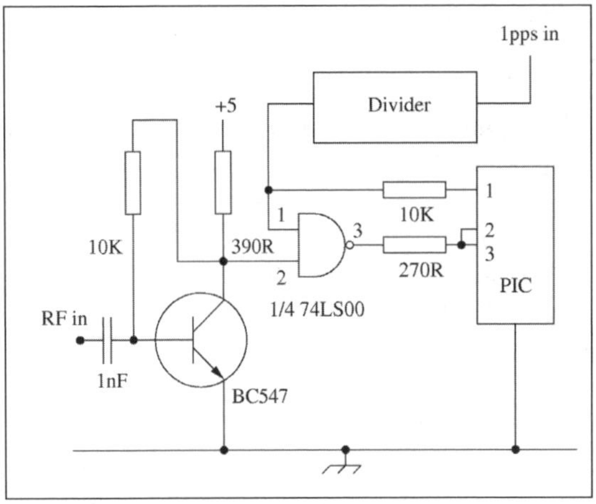

FIGURE 3.16: Simplified circuit of the counter.

stations and may be out of date. The ephemeris data which is used to indicate the satellites position is not always absolutely accurate and can be as much as four hours old before it is considered invalid. The propagation time through the Earth's atmosphere is subject to variations caused by changes in the ionosphere. Multi-path reception caused by signals reflected from nearby objects will also degrade the accuracy of the system. Despite these problems, a typical consumer grade GPS receiver can keep accurate time to within a few tens of ns. Before March 2000, the standard GPS navigation system included deliberate errors which were used to limit the accuracy of the system for most users. The SA (selective availability) feature was disabled on 1 March 2000 **[32]**. This resulted in a significant improvement in the accuracy of the system for civilian users.

Most GPS receiver modules have a 1pps (one pulse per second) output which produces a 1 second marker which is aligned with the beginning of the UTC second. Many GPS receiver manufacturers specify a maximum error of ±1 microsecond for the 1 second marker signal. Since SA was disabled, even cheap consumer grade receivers can usually manage to achieve better than ±100ns.

GPS CONTROLLED COUNTER. I have acquired no less than three GPS receivers during the last year. The first receiver is a Garmin GPS-25 OEM (original equipment manufacturer) module. This receiver was purchased on eBay at a price of about £14. The second receiver is a Rockwell Jupiter 8 which was also bought on eBay for a bargain price of £10. The third receiver is a Maplin A09FN Bluetooth GPS receiver which cost £40 under special offer. The Garmin and Rockwell units are OEM modules, which means that they come as a basic PCB module without a display, power supply, aerial or any other support hardware. These units are normally used in vehicle tracking systems, aircraft, trains etc. The Maplin receiver is a small battery powered unit which uses wireless Bluetooth technology to send time and navigation information to a Bluetooth equipped computer, PDA or mobile phone. I decided to build a frequency counter which uses the GPS 1pps signal to control the counter gate. This counter will be used to evaluate the performance of the GPS receivers. This turned out to be a much easier project than expected. I used that trusty old workhorse, our original frequency counter to do the counting. The GPS 1pps signal was divided and turned into a 50% duty cycle square-wave using a couple of 74LS390 logic ICs. This allows a number of particularly useful division ratios including 2, 20, 200 and 2000. Note that the counter gate is only open during the positive period of the divider output which gives counter gate times of 1s, 10s, 100s and 1000s. A standard two input NAND logic gate is used as the counter gate. The transistor amplifier is similar to the circuit used in the original counter project. **Figure 3.16** shows a simplified schematic of the circuit. As the gate time is determined by the GPS receiver, the frequency of the PIC's clock oscillator is not critical. Any old crystal of around 10MHz +/-10% should do the job.

TESTING THE MAPLIN UNIT. The first receiver tested was the Maplin unit. When I bought this receiver, I knew it was based on the SiRFstar III chipset which has a very good reputation for accuracy, even when it is used indoors where the GPS signals are very weak. I had no way of telling whether it had an internal 1pps signal which would be accessible on the PCB. Before we proceed any further, I will issue my usual disclaimer: Dismantling a brand new GPS receiver and poking a hot soldering iron into it will almost certainly invalidate the warranty. If you do happen to let the smoke out of it, it would not be reasonable to expect the vendor or manufacturer to repair or replace it under warranty. You have been warned!

The GPS receiver module inside the Maplin A09FN is a small tin-can which is surface mounted on the top of the main PCB. Initial tests with a multimeter did not look very promising. There was no sign of anything that looked even remotely like a 1pps signal on any of the receiver module pins. A quick test with an oscilloscope didn't reveal anything either. Finally, I tried connecting the oscilloscope probe to the trigger input of the scope and setting the X timebase switch to external triggering. When the probe was connected to one of the pins on the GPS module, the scope trace gave a very quick flash once every second. The pulses on this pin seemed to be exactly aligned with the RWM one second pips. It seems that this receiver produces a very narrow positive pulse of 1μs duration or perhaps even less at the beginning of each UTC second. This signal was used to drive the frequency divider in the counter circuit. The 3.3volt logic level 1pps signal from the GPS unit proved capable of driving the 5volt 74LS390 counter/divider. **Photo 3.17** shows details of where to find the 1pps pin in the Maplin receiver. The 1pps signal is taken via a 68Ω resistor from the second pin in from the corner of the GPS module nearest to R25. The 68Ω resistor is only intended to act as a convenient mounting point for the wire to the external counter/divider. The ground connection of the 1pps signal connects to the GPS receiver at the grounded side of C7 which can be seen on the left side of the photo.

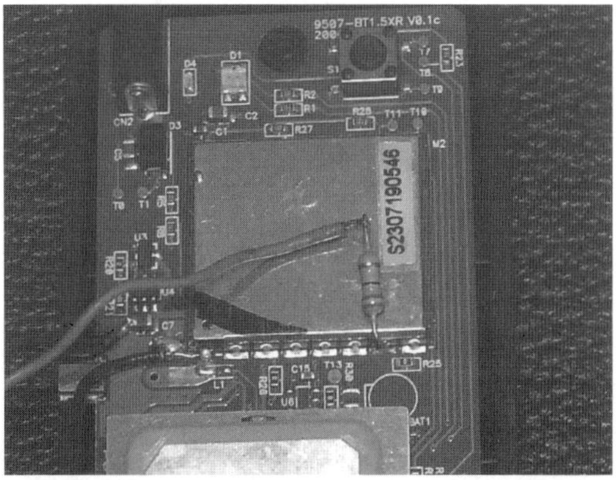

PHOTO 3.17: Location of the 1pps pin in the Maplin GPS receiver.

The signal from my 10MHz OCXO was fed to the frequency counter input. As the output from this oscillator is known to have extremely good short term stability, we can assume that any sudden changes in the count result are caused by jitter in the GPS derived gate signal. It is not necessary for the

OCXO output frequency to be very accurate for this test. At this point, we need stability rather than accuracy. The OCXO has a warm up time of only about 3-5 minutes, but I let it run for one hour before doing any tests so that I could be sure it was fully warmed up and stable. With a gate time of one second, the indicated cycle count was 10,000,000 cycles. I would normally say plus or minus one cycle at this point, but in fact, the count result was +1 / -0 even over a period of several hundred seconds, indicating that the GPS 1pps jitter is indeed less than ±100ns. This is an error of less than 1 part in 10 million per second of gate time. $1x10^{-7}$ for a 1 second gate, $1x10^{-8}$ for a 10 second gate and $1x10^{-9}$ for a 100 second gate time. This is exactly the sort of accuracy that is required to calibrate our OCXO frequency standard.

Similar tests were performed using the Rockwell and Garmin receiver modules. Finding the 1pps marker signal outputs on these receivers was just a simple matter of consulting the manufacturers data-sheets. All three GPS receivers showed 1pps accuracy which was consistently better than ±100ns. The counter input frequency was increased to 30MHz by placing a frequency tripler between the OCXO and the counter input. Again, all three counters performed well in this test.

RECEPTION. The Rockwell and Garmin receivers were fitted with an external active aerial which was mounted on the shack roof with a reasonably good view of the sky. The Maplin unit uses an internal 'patch antenna' which was used indoors in my windowless shack. Despite this huge disadvantage, the Maplin receiver performed as well as or better than the other two receivers in every test. During a test period of several thousand seconds, the Maplin unit 1pps jitter was always less than ±2 cycles of the 30MHz signal (66ns) and for 98% of the total test time was no more than ±1 cycle (33ns). None of the receivers showed 1pps jitter in excess of 100ns at any time during the tests, not even for a single one second cycle. The status of the GPS receivers was monitored using gpsd, xgps **[33]** and gpsdrive **[34]** software running on a Linux PC. The receivers with a proper external aerial were tracking between 6 and 11 satellites and showing a low PDOP (position dilution of precision) reading of 2 or less for most of the tests. The SiRFstar III based Maplin receiver was struggling to receive more than 3-4 satellites and showing a relatively high PDOP figure of 4-8 which makes its good performance all the more remarkable. There is one slightly annoying problem with the Maplin unit. It will automatically power down after a few minutes if it is not connected to a Bluetooth device. This is not a big issue for me because my PC runs 24/7 anyway. If there is a way of disabling this feature, I have so far been unable to find it.

The GPS controlled counter has turned out to be a very powerful tool for frequency measurement and for calibrating OCXOs. Gate times of 10s or longer are easily selected by reconfiguring the 74LS390 dividers. It would be a simple matter to replace the decade counter/divider ICs with a binary counter like the 74HC4040. This would allow gate times of

64, 128 or 256 seconds which would give a resolution of 1Hz when the counter is used with a UHF prescaler as used in last month's counter project. It would not be difficult to configure the PIC to provide a correcting voltage to the OCXO EFC (electronic frequency control) input to make a GPS disciplined oscillator. This approach was used in the VE2ZAZ GPS derived frequency standard **[35]**.

The construction of the counter is very straightforward. The 74LS390s and the 74LS00 are mounted dead-bug style on a sheet of PCB laminate. Most of the work is done by the PIC counter. Hacked firmware which allows the counter to be used with a GPS derived gate signal is available on the web: http://tinyurl.com/q24xj.

GPSDO. This next project is a GPS disciplined 10MHz frequency standard. I built two versions of this project. The first version used a 10 bit binary counter to divide the 10.240MHz clock signal of the original PIC counter project down to 10kHz so that it could be phase locked to the 10kHz timing signal from the Rockwell Jupiter receiver. I didn't have a suitable 10 bit counter available so I used a pair of 74LS393 8 bit counters instead. An exclusive-or gate was used as the phase detector and a varicap diode was used to control the frequency of the PIC's 10.24MHz crystal. I won't describe the circuit in detail here because it is almost identical to the 10MHz GPSDO described below.

The second version of the GPSDO is based on a simple 10MHz crystal oscillator. The 10MHz signal from the oscillator is buffered by a simple source-follower transistor amplifier. One of the NAND gates of the 74LS00 IC is used as a second buffer amplifier. The high level 10MHz signal is then divided by 1000 using a pair of 74LS390 dual decimal counter ICs. This results in a 10kHz square-wave signal which is compared with the 10kHz timing signal from the Rockwell Jupiter 8 GPS receiver using a 74LS86 exclusive-or gate as a phase detector. The output of the phase detector is low pass filtered to remove the 10kHz reference signal and any other high frequency components. The resulting DC voltage is applied to a varicap diode in the 10MHz oscillator.

CONSTRUCTION. The schematic of the crystal oscillator is shown in **Figure 3.17**. This oscillator needs to have a maximum free-running frequency drift

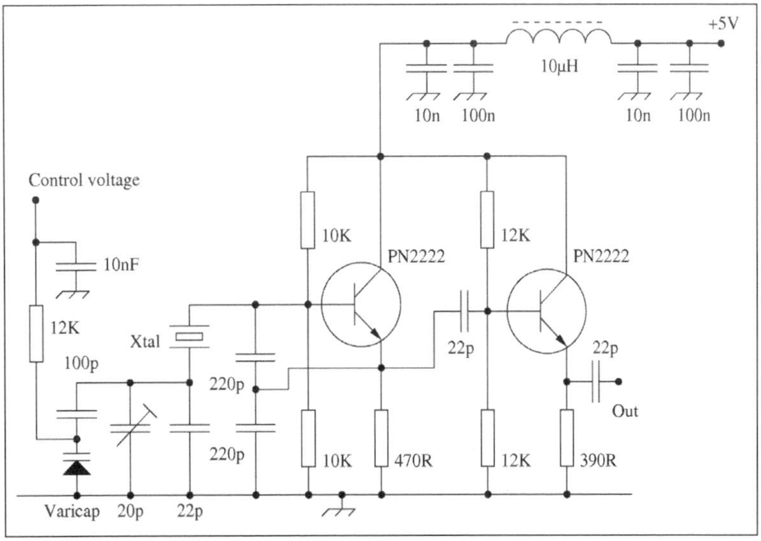

FIGURE 3.17: Circuit of the crystal oscillator.

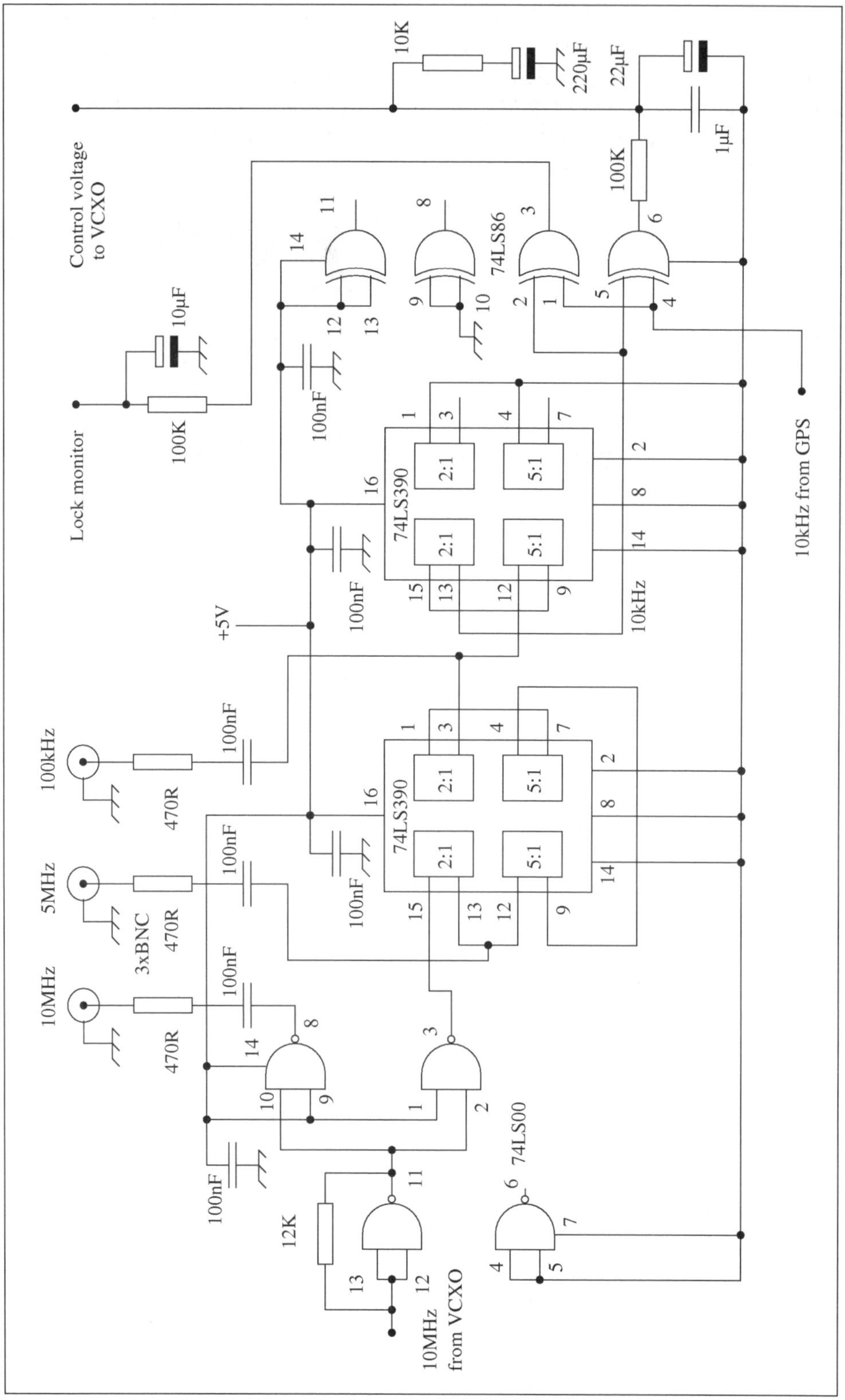

FIGURE 3.18: Circuit of the crystal oscillator.

of no more than about 3Hz/hour in the first hour after a five minute warm up period and less than 1Hz/hour after a warm up time of several hours. The 10MHz crystal should be of the highest possible quality. The capacitors in the oscillator should be NPO ceramic or polystyrene types. The frequency divider and phase detector were built dead-bug style on PCB laminate. **Photo 3.18** shows the layout of the circuit. The schematic is shown in **Figure 3.18**. I used a BB109 varicap diode in the original frequency control circuit. Since varicap diodes seem to be quite difficult to find these days, the BB109 was later replaced by a standard red LED **[36]**. Note that the LED should be reverse biased for use in this circuit. The finished oscillator should give about 50Hz frequency change per volt of applied DC control voltage.

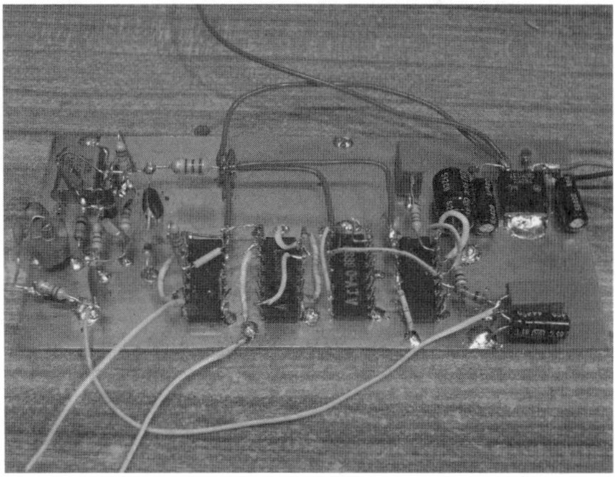

PHOTO 3.18: Layout of the crystal oscillator.

Figure 3.19 shows the schematic of a small interface board. One of the two transistors in this unit is used as a buffer/inverter for the TXD (transmit data) signal from the GPS receiver. The output from this circuit is a 5V logic level signal rather than the ±11 volts produced by a typical PC serial port. I have found that this 5V output is capable of driving the serial port input of my desktop and laptop PCs. If you need a proper RS232 output, this circuit can be replaced by a MAX232 IC. The second transistor acts as a buffer between the 1pps signal from the GPS and a LED mounted on the front panel of the enclosure. The flashing LED acts as a 'heartbeat' to show that the GPS receiver is running.

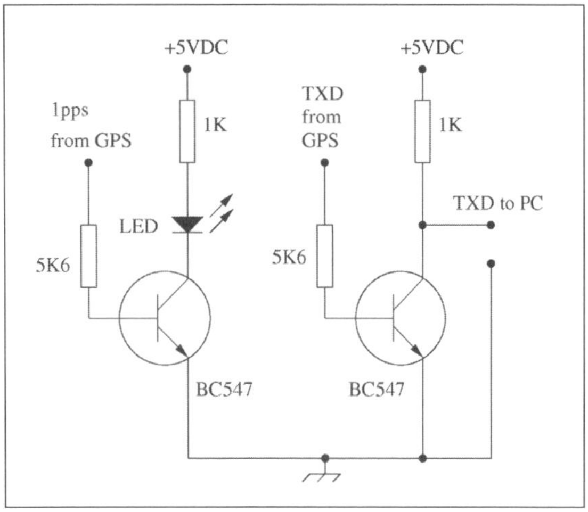

FIGURE 3.19: Circuit of small interface board.

The GPSDO requires a stable 5V power supply. I used a simple mains power supply as shown in **Figure 3.20 (overleaf)**. The mains signal is filtered by an IEC connector with a built in RFI filter. A fuse is fitted to the live side of the mains connection. I used a 100mA fuse which was the smallest value available. A 230V to 0-6 – 0-6 volt RMS transformer is used to produce 12VAC which is rectified by a bridge rectifier made from four 1N4004 1A rectifier diodes. The resulting 17VDC (actually about 15V under load) is reduced to 8V by a 7808 regulator and finally to 5V by a 7805. Both regulators should be bolted to the bottom of the aluminium enclosure

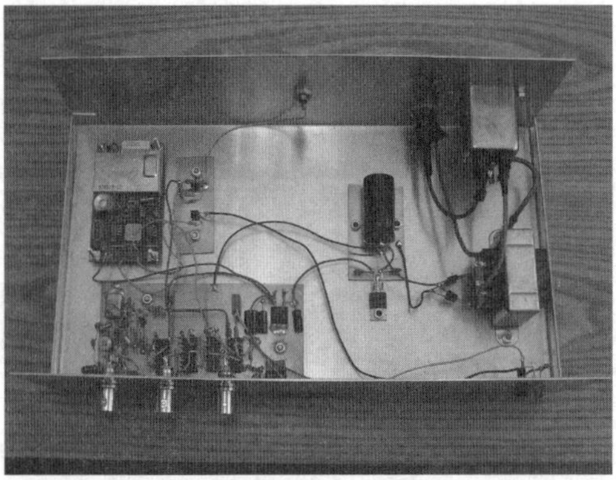

PHOTO 3.19: Inside the GPSDO unit.

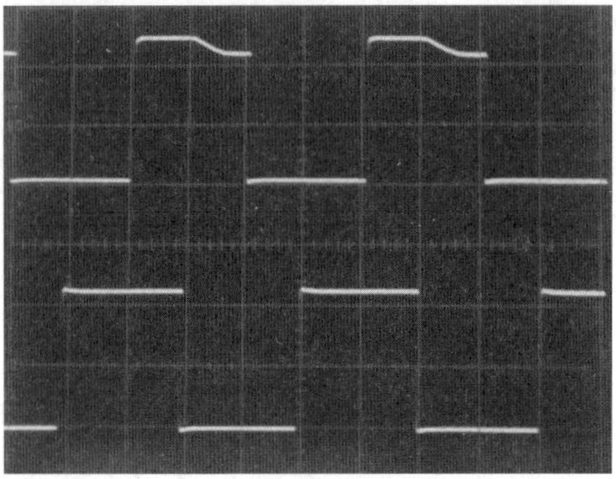

PHOTO 3.20: 10kHz signals applied to the phase detector.

or a suitable heatsink. The low voltage side of the mains transformer is protected by a 1A fuse. **Photo 3.19** shows the inside of the GPSDO unit. The enclosure is a Maplin instrument case. The GPS and interface board are at the rear of the enclosure on the left. The oscillator and PLL circuit are at the front left and the power supply is on the right.

TESTING. I used the Temex OCXO which has been calibrated accurately to within 1ppb (±0.01Hz) using the GPS controlled counter described earlier as a reference for checking the stability of the GPSDO output signal. I found that the frequency of the GPSDO is accurate to within 1×10^{-7} within five minutes of powering up the unit, 1×10^{-8} with 7 minutes and 1×10^{-9} within 30 minutes. At the time of writing, the GPSDO has been running for five days. Monitoring the two 10MHz signals with a dual channel oscilloscope shows that the free running OCXO is drifting at a rate of less than one cycle in several minutes! The 10kHz signals applied to the phase detector are shown in **Photo 3.20**. Note the phase difference of approximately 90 degrees when the loop is locked. The finished GPSDO is shown in **Photo 3.21**. One of the exclusive-or gates in the 74LS86 is used as a simple lock detector. The output from this lock monitor is available from a connector on the front panel. The voltage at this point will be approximately the same as the voltage applied to

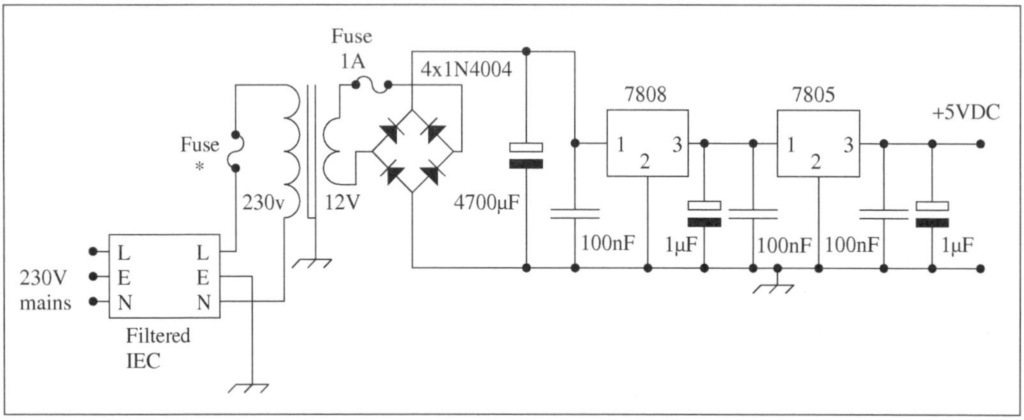

FIGURE 3.20: Simple mains power supply for the GPSDO.

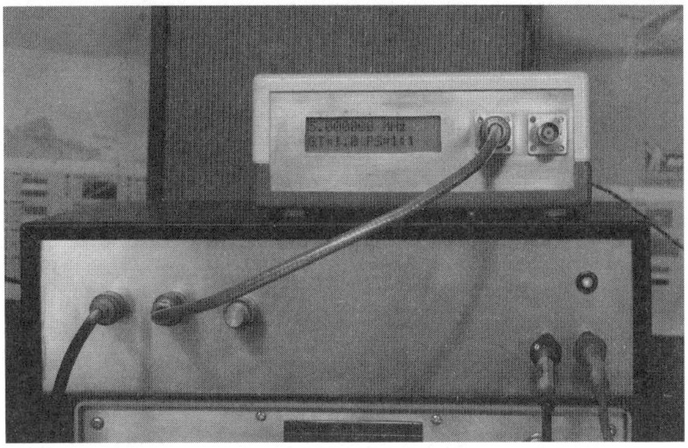

PHOTO 3.21: The finished GPSDO.

the varicap diode. You can check if the loop is locked by measuring the voltage at the lock monitor output with a digital voltmeter. The loop is locked if this voltage remains stable to within ±1 percent over a period of a few minutes. The trimmer capacitor in the oscillator should be adjusted so that the DC voltage at the lock monitor output is about 2.0 volts. This adjustment is quite tricky to get right, but hopefully you will only need to do it once.

FINDING A SUITABLE GPS. Rockwell Jupiter GPS units are often available on eBay. There are two Jupiter 8 units on auction at the time of writing. I have seen at least a dozen units for sale in the last couple of months. The Jupiter-T is ideal for use in a GPSDO, but they are not so easy to find. The Jupiter 8 is easier to find and usually sells for £10-£20. Later models with firmware version 1.8 are recommended, but I have found that my version 1.5 unit manufactured in late 1997 is stable and well behaved. I have read several reports that earlier versions of the Jupiter receiver are not stable enough for use in a GPSDO. Check the firmware version with the seller before bidding. Don't be tempted to pay more than about £20 even for a later model. Modern state-of-the-art GPS receivers that are designed specifically for timing applications like the Navsync CW12-TIM **[37]** cost about £50-£70 for a new unit with a one year guarantee. The CW12-TIM has a built in NCO (numerically controlled oscillator) with a default output frequency of 10MHz, which makes a homebrew GPSDO project rather too easy!

WEBSEARCH

http://homepage.eircom.net/~ei9gq/pic.html

Lots of PIC resources including links to all the software mentioned in this chapter

RECOMMENDED READING:

'A GPS-Based Frequency Standard'. Brooks Shera, W5OJM. *QST* July 1998. 'VE2ZAZ GPSDO' see [35]

'Frequency Measurement for Experts'. M. Greenman ZL1BPU. http://www.qsl.net/zl1bpu/PRECISION/freqexpert.htm

'Probably the Simplest GPS Disciplined Oscillator possible !' - A. Talbot G4JNT. http://uk.geocities.com/ac.talbot@btinternet.com/SimpleGPSDO.pdf

References
[21] http://homepage.eircom.net/~ei9gq/iambic.html
[22] http://ww1.microchip.com/downloads/en/DeviceDoc/33014J.pdf
[23] http://ww1.microchip.com/downloads/en/DeviceDoc/40044D.pdf
[24] http://www.microchip.com Search for: AN592
[25] Embedded Control Handbook 1994/95 AN544
[26] www.maplin.co.uk/Media/PDFs/N27AZ.pdf
[27] http://www.npl.co.uk
[28] http://www.g4jnt.com
[29] http://www.jrmiller.demon.co.uk/
[30] http://www.xs4all.nl/~barendh/Indexeng.htm
[31] http://www.iz8ede.it/datasheets.htm
[32] http://en.wikipedia.org/wiki/GPS
[33] http://gpsd.berlios.de/
[34] www.gpsdrive.de/
[35] A Simplified GPS Derived Frequency Standard. B Zauhar VE2ZAZ. *QEX* Sep/Oct 2006.
[36] www.hanssummers.com/radio/varicap/index.htm
[37] www.navsync.com

4

Transmitting Projects

RECYCLING. By re-using some of the modules from previous projects you can build an SSB exciter. SSB (single sideband) is a form of AM (amplitude modulation). **Figure 4.1A** shows the spectrum of a full carrier AM signal where the carrier is modulated by a 1kHz tone. The carrier frequency is 3.700MHz. The upper and lower sideband signals are at ±1kHz. If the modulation level is 100%, the power in either sideband is 25% of the carrier power. **Figure 4.1B** shows the spectrum of a voice-modulated AM transmitter. If a low pass filter is used to limit the bandwidth of the speech signal to a maximum of 3kHz, the transmitted AM signal will have a minimum bandwidth of 6kHz. This assumes that we are using an ideal AM transmitter that generates no distortion or noise. A real transmitter will always produce unwanted distortion that will increase the bandwidth of the transmitted signal.

Full carrier AM is a very inefficient way of transmitting a voice signal. Even at 100% modulation, the carrier contains at least two thirds of the transmitter power. As the carrier contains no useful information, it can be suppressed at the transmitter. An AM signal with a suppressed carrier is known as double sideband, suppressed carrier (DSBSC). Suppression of the carrier leads to a significant reduction in power consumption.

A DSBSC signal can be generated by mixing the RF carrier with the modulating audio in a balanced or double-balanced mixer. There is little or no difference between a balanced mixer that is used in a receiver and a mixer that is used to generate a DSB signal in a transmitter. A balanced mixer that is used as a modulator is known, not surprisingly, as a balanced modulator. Single-balanced diode mixers like the one used in the direct conversion receiver and double balanced mixers (DBMs) used in the 80m receiver project in chapter 2 make excellent balanced modulators.

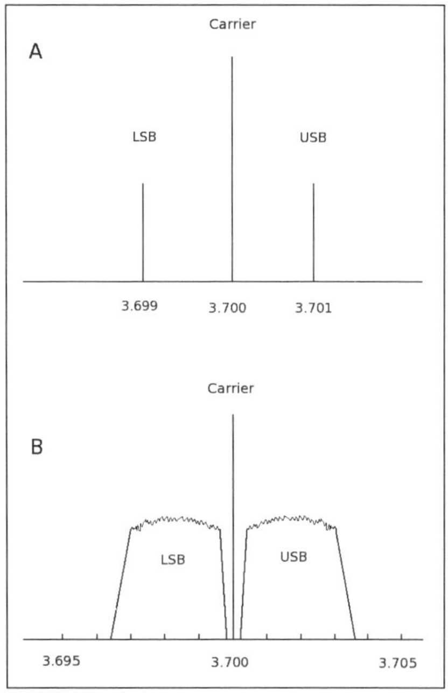

FIGURE 4.1:
A: The spectrum of a full carrier AM signal where the 3.7000MHz carrier frequency is modulated by a 1kHz tone.
B: The spectrum of a voice-modulated AM transmitter

Despite its low cost and extreme simplicity, the simple two-diode balanced modulator (**Figure 4.2**) is capable of very good performance. Diode DBMs make excellent balanced modulators. I have avoided using diode DBMs as balanced modulators in the past because they don't usually provide any method of adjusting the carrier-balance. Commercially made DBMs like the SBL-1 have excellent port-to-port isolation so a balance control is not really necessary. When I was experimenting with the diode mixer for the front-end of the 80m superhet receiver, I discovered that it is quite easy to build a DBM with port-to-port isolation well in excess of 50dB. **Figure 4.3** shows a diode DBM used as a balanced modulator.

As each sideband of our DSBSC signal is a perfect mirror image of the other sideband, all of the modulation information is contained in both sidebands. We can make further savings in power consumption by removing one sideband. The unwanted sideband can be removed by a very selective IF filter like the crystal ladder filter from the 80m receiver project. As well as improving the transmitter efficiency, removal of the unwanted sideband reduces the bandwidth of the transmitted signal by more than 50%.

PHASING METHOD. There is another popular method of generating an SSB signal. Many of the earlier SSB transmitters used a phasing system to cancel out the opposite sideband. This method uses a pair of balanced modulators. The carrier signal is fed to the two mixers in phase-quadrature which means that one carrier signal is 90° out of phase with the other. The audio signal is also applied to the two balanced modulators in phase quadrature. The DSB output from the two modulators is summed (added) in a way that reinforces the wanted sideband and cancels out the unwanted sideband. This type of phasing system is known as a Hartley Modulator. It is relatively easy to generate the phase-shifted carrier signals with logic ICs or RC phase shift networks. It is not so easy to generate the phase-shifted audio signals. An SSB speech transmitter will have an audio bandwidth of somewhere between two and three kHz. A typical example would have an audio bandwidth of 2.7kHz ranging from 300Hz to 3kHz.

Designing a phase shift network that will maintain its accuracy over this 10:1 frequency range is not a trivial exercise. Many of the early SSB phasing rigs used a simple RC audio phase shift network.

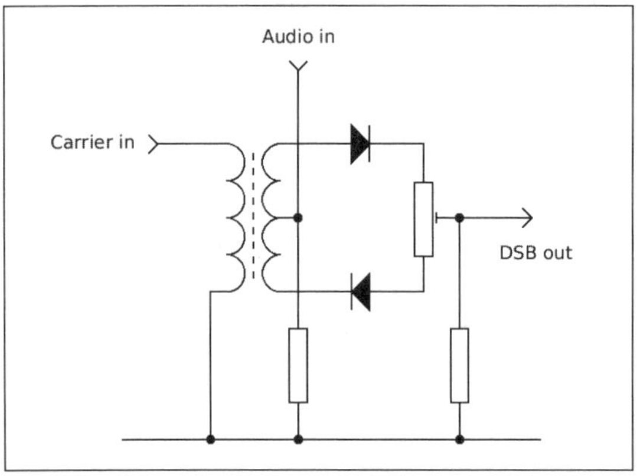

FIGURE 4.2: A simple two-diode balanced modulator.

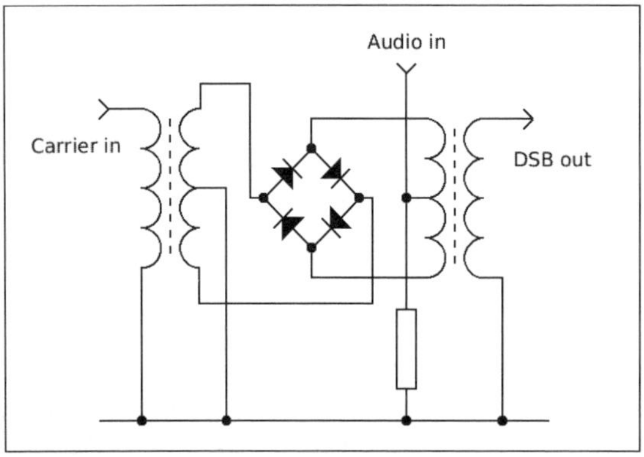

FIGURE 4.3: A diode DBM used as a balanced modulator.

These rigs often (always?) suffered from poor opposite sideband suppression. Due to the availability of high quality 1% tolerance components and active phase-shift networks using op-amps, it is now possible to build high performance phasing rigs using readily available components. **Figure 4.4** shows a simplified block diagram of a phasing exciter.

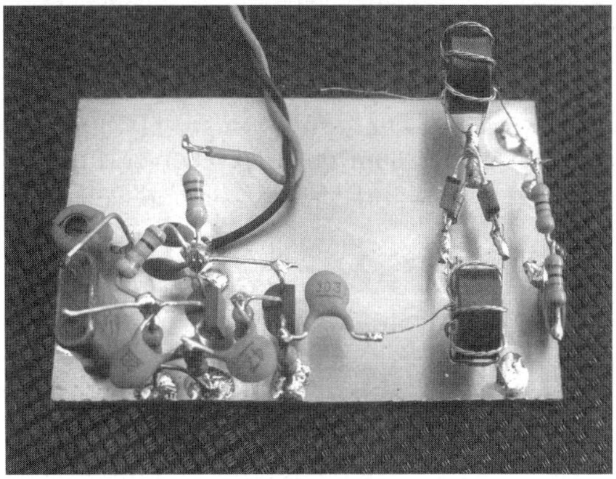

PHOTO 4.1: Eamon's dead-bug version of the oscillator and modulator.

A small and inexpensive DSP system can be configured to act as a very high performance phasing transceiver. The use of software algorithms to perform the audio phase shifts makes it possible to build a DSP phasing system that would be difficult or perhaps even impossible to build using analogue phase shift networks. There are a few variations on the phasing method of SSB generation. A slightly more complex system known as the Weaver Method or Third Method is quite popular with radio amateurs. Another phasing system is known as the Fourth Method **[38]**. Since I have never built any equipment using these newer methods, I won't discuss them any further.

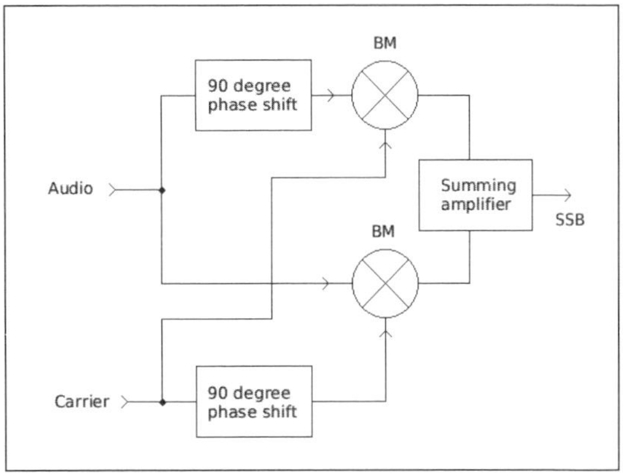

FIGURE 4.4: A block diagram of a phasing exciter.

THE CIRCUIT. Drawing the schematics for this project was a very easy job. Each stage of the SSB exciter was borrowed from one of the many circuits in this book.

Figure 4.5 shows the schematic of the oscillator/modulator. I built the oscillator and modulator stages on a scrap of PCB using dead-bug style construction. The construction details and some rather dodgy soldering can be seen in **Photo 4.1**. The microphone amplifier is on a separate PCB. If you haven't build the microphone amplifier project, it can be replaced by any low to medium output impedance audio amplifier.

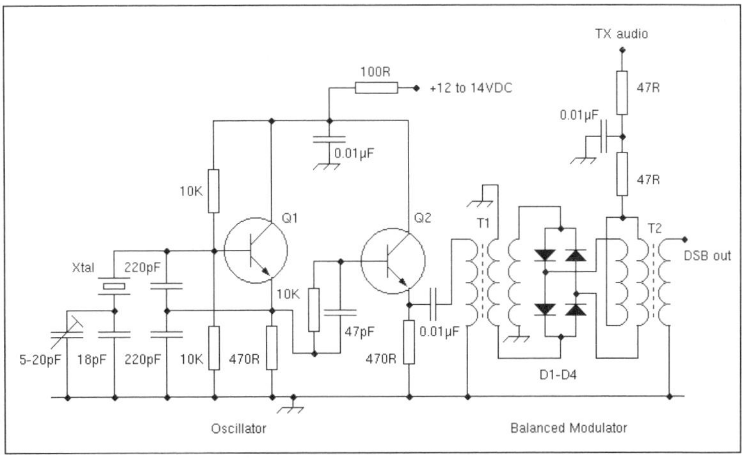

FIGURE 4.5: The schematic of the oscillator/modulator.

The construction details are quite straightforward – the output from the oscillator unit is connected to the local oscillator port of the DBM. The transmit audio is fed to the IF port of the DBM via a simple diplexer/ LPF. The double-sideband output is taken from the RF port of the DBM. I used a ring of four BAT43 Schottky diodes for D1-D4. The diodes are all from the same batch, but no attempt was made to match them for forward resistance. T1 and T2 are made from six turns, trifilar wound on ferrite toroids. FT37-43 or similar ferrite cores should be suitable. I used BC548 transistors instead of the 2SC945 used in the original circuit.

You could use a balanced mixer IC for the modulator instead of the diode mixer/modulator. RF mixer ICs like the Signetics/Philips 602/ 612 and the obsolete but still popular Plessey 600 series are low-level mixers designed for minimum power consumption rather than best linearity and maximum dynamic range. It is worth noting that the cheap and simple diode modulators described will give much better performance than the average IC mixer. Even the best and most expensive IC mixers will have difficulty matching the performance of our simple diode modulator.

TESTING. As usual, only the most basic test equipment was used for testing the SSB exciter. I connected a short length of wire to the DBM's RF port for the tests. A homemade receiver was used to monitor the signal radiated from this wire. A 3.579MHz crystal was used in the carrier oscillator so that the DSBSC signal would fall within the 80m band. I used a 51Ω resistor and a diode probe to test the output of the oscillator module. This test was performed before the oscillator module was connected to the DBM. The oscillator is capable of delivering more than 1V RMS into a 51Ω load (+13dBm).

I always follow a simple rule-of-thumb for setting up balanced modulators. The maximum audio power applied to the modulator should be around 20dB lower than the carrier power. This ensures that Inter-modulation distortion (IMD) products are kept to a minimum. This means that we need about 100mV RMS of audio at the IF port of the DBM. The microphone amplifier gain is fixed at 20dB so any handheld or desk microphone capable of delivering about 10mV into 10kΩ should be suitable for this job. If necessary, the gain of the microphone amplifier can be adjusted by changing the value of R5 or replacing it with a 1MΩ pot.

After the carrier oscillator and microphone amplifier were connected to the DBM, I tuned my receiver to 3.579MHz and listened to the DSB signals from the exciter. The first results were very encouraging, the quality of the DSB signal seems quite good and the carrier is well suppressed. You can listen to a recording of this test at: http://tinyurl.com/ eqbc8. Despite the relatively strong (59+) signal, I couldn't find any traces of spatter beyond ±3kHz from the signal frequency. Remember that this simple transmitter has no crystal filter to suppress unwanted

IMD products. It would seem that this simple gadget is quite capable of generating very high quality DSBSC signals. To generate SSB, we only need to add a crystal filter and change the carrier crystal.

I used an RF millivolt-meter to measure the output at the RF port of the DBM. With no audio applied to the microphone amplifier, the RF output is just under 0.5mV which is 66dB below the carrier level. With normal audio levels from the microphone amplifier, the peak DSB output is around 100mV. A good single-tone whistle will drive the output to 300mV RMS. This means that the carrier leakage at the output of the modulator will be around 45-50dB below the peak output. As the IF filter will add another 15-20dB of carrier suppression, it should be possible to achieve carrier suppression in excess of 60dB.

For the final test, I put a crystal SSB filter between the output of the DBM and changed the carrier frequency to 10.6975MHz by replacing the crystal. The SSB filter has a passband centre frequency of 10.695MHz. The test confirmed that carrier leakage was well below -60dB relative to the PEP output. I was not able to monitor the SSB signal because I don't have a receiver that covers 10.695MHz. We can expect the balanced modulator to have a single-sideband conversion loss of around 6dB. The crystal filter will lose a few more dB. This means that the SSB peak output will be no more than about 20-30mV.

PRACTICAL APPLICATION. The SSB exciter will be used as part of an SSB transmitter or transceiver. **Figure 4.6** shows a block diagram of a complete SSB transmitter. Our simple DSBSC generator is shown within the dashed lines. The local-oscillator, IF filter, RF BPF and various buffer amplifiers have been covered in the 80m RX project. The only missing blocks are the driver, PA and transmit low-pass-filter.

Most SSB equipment has some means of shifting the carrier oscillator frequency for switching between upper and lower sideband mode. To generate a lower sideband (LSB) signal, the carrier oscillator frequency must be adjusted so that is just above the passband of the IF filter. This will allow the lower sideband to pass through the filter. The upper sideband will be removed by the filter. For USB, the carrier frequency is set just below the filter passband.

There are two commonly used methods of switching the carrier frequency. You can switch an L or C reactance in the oscillator circuit that will

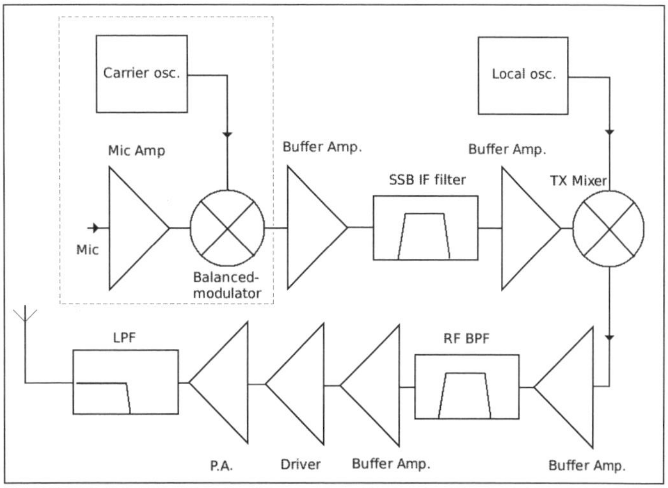

FIGURE 4.6: A block diagram of a complete SSB transmitter.

pull the oscillator frequency to the desired value. The usual arrangement has the oscillator frequency on the high side of the filter passband (LSB). An electronic switch is used to put an inductor in series with the crystal. This will pull the oscillator to a lower frequency on the low side of the filter passband (USB). The switch is usually a transistor or PIN diode. The big advantage with this method is that only one expensive crystal is required.

My preferred way of switching sidebands is to use a pair of oscillators and simply switch the power supply to the correct oscillator for the required mode. **Figure 4.7** shows a schematic of the carrier oscillator used in my HF transceiver. **Figure 4.8** shows the complete exciter section of this transceiver. Note that the buffer amplifier for the carrier oscillator has an output for the receive product detector so that the same carrier oscillator can be used for both transmit and receive. T1 is an IF transformer with a centre-tapped secondary winding. This transformer is identical to the one used in the second IF stage of the 80m receiver. The 2.2kΩ resistor at the DSB output is used to provide a few mA of DC to a diode switch at the input of the IF filter. This resistor can be removed if you don't need it.

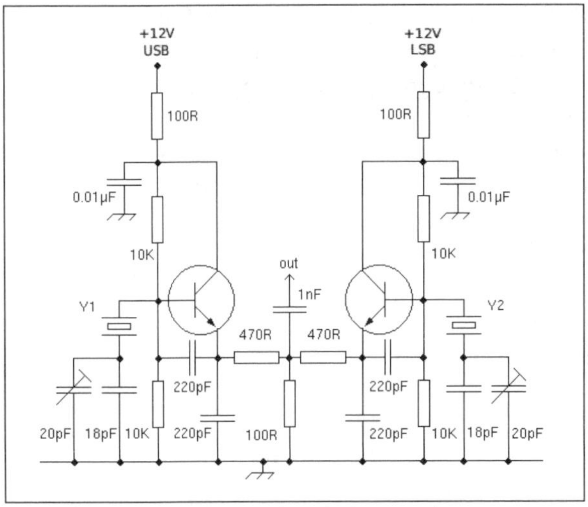

FIGURE 4.7: A schematic of the carrier oscillator used in my HF transceiver.

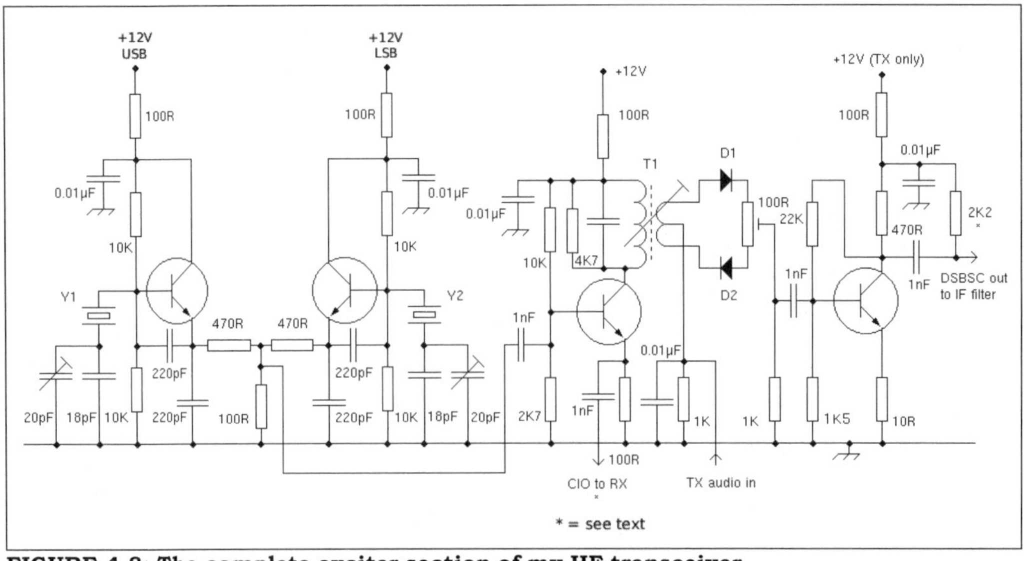

FIGURE 4.8: The complete exciter section of my HF transceiver.

Linear Amplifiers and an Amplifier for an SSB Transmitter

AMPLIFIERS. Apart from the much greater power levels involved, the requirements for a good transmit RF amplifier are much the same as for a receive RF amplifier. See the 80m receiver project for detailed information about linearity and two-tone dynamic range. Discrete transistors or ICs can be used as the active devices. Bipolar transistors, JFETs, MOSFETs and valves are all widely used as transmit amplifiers. All of the amplifiers described here will use bipolar transistors in the common-emitter configuration.

Some modulation modes do not require linear amplification. Transmitters for FM (frequency modulation) and the various other kinds of phase/frequency/angle modulation can use non linear Class C amplifiers. AM transmitters that use high-level modulation of the final stage can use Class C amplifiers. CW transmitters are also included in this category, provided that the keying is done in the final stages.

OPERATING CLASS. A Class A transistor amplifier is biased so that the transistor is conducting for all of the input signal cycle. **Figure 4.9A** shows a simplified Class A amplifier. The input drive level should always be kept low enough to ensure that the output signal voltage is always more than zero and the peak signal current is always less than the standing current. Because of the relatively high DC standing current, Class A amplifiers are very inefficient. The theoretical maximum efficiency is 50%. In practice, most Class A amplifiers will have an efficiency of less than 25%. This is not a big problem for low-level amplifiers. High power linear amplifiers are usually operated in more efficient modes like Class AB. Some commercial equipment has a Class A mode that allows the operator to select Class AB for normal operation at high power levels or Class A mode for improved linearity at reduced power levels.

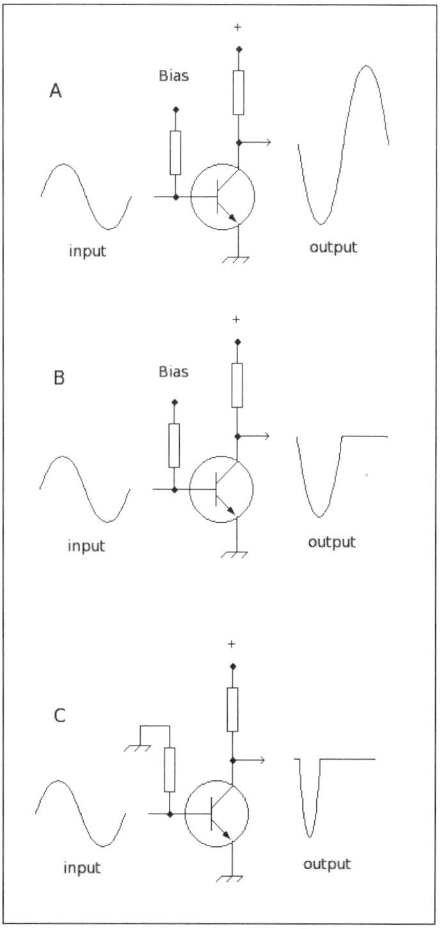

FIGURE 4.9: The input and output from Class A, B and C amplifiers.

Class B amplifiers are biased so that the DC standing current is zero. **Figure 4.9B** shows a Class B amplifier. The transistor only conducts on positive half-cycles of the input signal. The problem of the missing half-cycles can be overcome by using a pair of transistors in a push-pull configuration. One transistor will conduct on negative half-cycles of the input signal; the other transistor will conduct on positive half cycles. **Figure 4.10** shows a push-pull transistor amplifier. Class B push-pull amplifiers are prone to a type of distortion that occurs at the crossover between positive and negative half-cycles. The distortion is caused by the voltage drop of the base-to-emitter junction and by the very non-linear characteristics of bipolar transistors when the collector current is close to zero. The 'crossover distortion' can be greatly reduced by biasing the amplifier for a small amount of standing current. This compromise between Class A and Class B operating conditions is known as Class AB. The efficiency of a typical Class AB RF power amplifier is around 50-60%. True Class B amplifiers are even more efficient, but they are rarely used as RF power amplifiers because Class AB offers a big improvement in linearity at a small cost in efficiency.

Class C amplifiers are biased below the zero-current cutoff point. The transistor will only conduct on positive peaks of the input signal. This means that the transistor will be biased off for more than 180° (>50%) of the input signal. Since silicon transistors require bias voltage of around 0.7V before any collector current flows, a transistor amplifier with 0V of bias will operate in Class C rather than Class B. It will be obvious from **Figure 4.9C** that Class C amplifiers are very non-linear. The amplified signal at the output is a much distorted replica of the input signal. The output of a Class C amplifier is also rich in harmonics. A tuned circuit or low-pass-filter at the output of a Class C amplifier will remove the harmonics and restore the RF waveform to something resembling a sine-wave. Class C amplifiers are capable of very high efficiency, typically better than 70%.

Other classes of operation like Class D use fast switching transistors that are switched by a square-wave signal so that the transistors are either biased completely off or completely saturated for most of the time. When such high-speed switches are used to drive an inductive load, the voltage across the transistor can fall close to zero when the transistor is on. The current is reduced to almost zero when the transistor is off. This means that power dissipation in the transistor will be kept to a minimum. Such amplifiers can be more than 90% efficient.

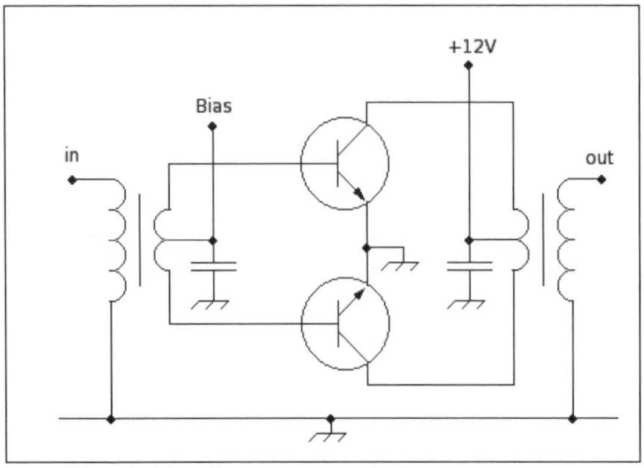

FIGURE 4.10: The schematic for a push-pull transistor amplifier.

NEGATIVE FEEDBACK. We can improve the linearity, stability and bandwidth of an amplifier by feeding some of the output signal back to the amplifier input. The feedback signal should be 180° out of phase with the input signal. **Figure 4.11A** shows a simple common-emitter transistor amplifier without any negative feedback. The amplifier in **Figure 4.11B** uses a resistor (Rf) to provide negative feedback from the collector of the transistor to the base. C is a DC blocking capacitor that passes the negative feedback signal without altering the DC bias conditions. Since a transistor in the common-emitter configuration is an inverting amplifier, the feedback signal will have the required 180° phase difference relative to the input signal. This type of negative feedback is known as shunt feedback. Another method of applying negative feedback is called emitter degeneration. If a resistor is placed in series with the transistor emitter, the voltage drop across this resistance will cause negative feedback.

Shunt feedback will reduce the input impedance of an amplifier. Emitter degeneration tends to increase the input impedance. By using a combination of shunt feedback and emitter degeneration, as seen in **Figure 4.11C**, it is possible to control the input and output impedance of an amplifier. The values of Rf and Re required for a given amount of gain and IO impedance will depend on the exact characteristics of the transistor and the other components used in the circuit. As a very rough rule-of-thumb, Rf=250Ω and Re = 10Ω will give just over 10dB of gain from a transistor amplifier with 50Ω I/O impedance. Rf=500Ω and Re=5Ω gives a gain of about 20dB **[39]**. Transistors for broadband RF amplifiers should have relatively high gain (hFE) and a high fT (unity-gain transition frequency). UHF power transistors like the 2N5109 and 2N3866 make excellent RF amplifiers.

The schematic shown in **Figure 4.12** is a screen capture of the QUCS (quite universal circuit simulator) **[40]** computer software. The

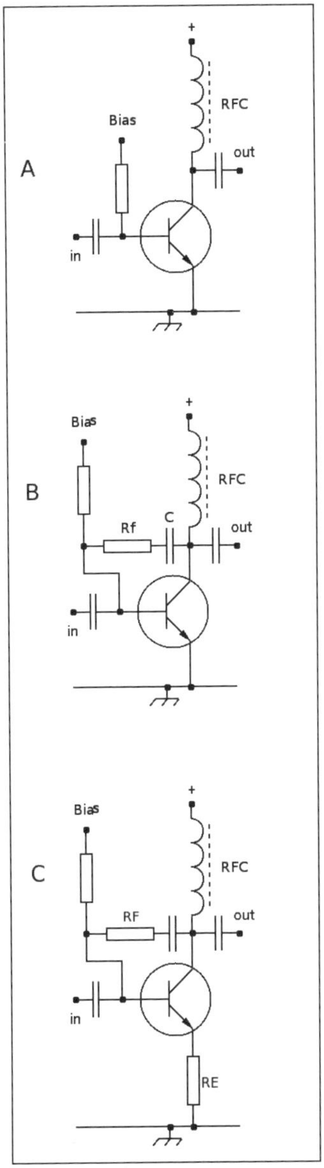

FIGURE 4.11: Schematics of basic amplifiers.

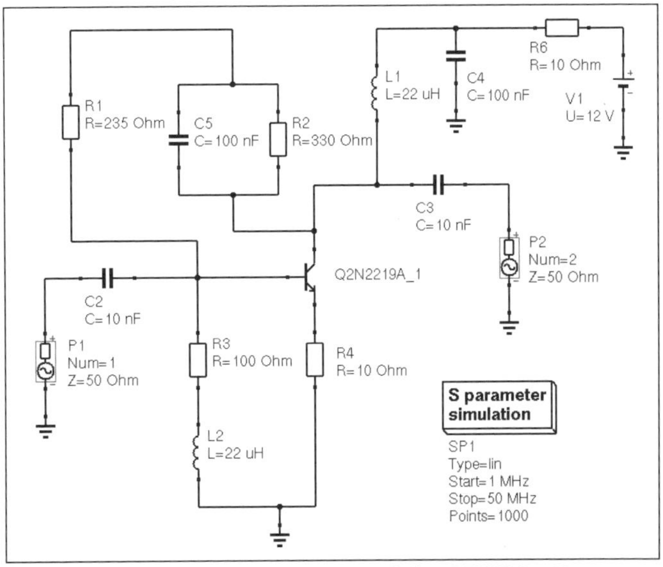

FIGURE 4.12: A screen capture of the QUCS (quite universal circuit simulator) computer software.

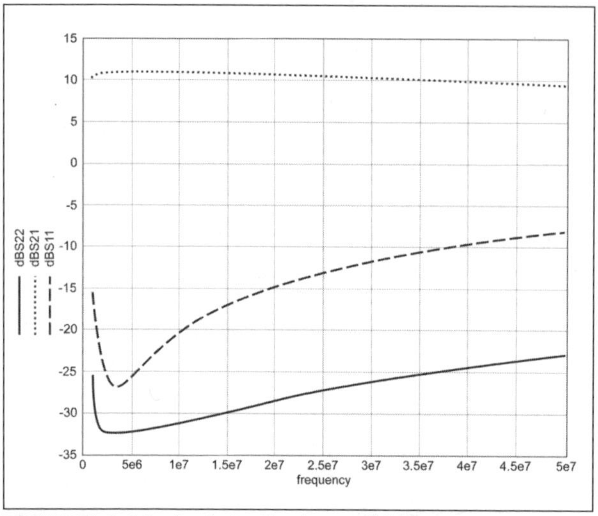

FIGURE 4.13: A plot of gain and I/O return loss for the amplifier shown in Figure 4.12.

amplifier gain is a little over 10dB. **Figure 4.13** shows a plot of gain and I/O return loss for this amplifier. The graph shows the amplifier's frequency response from 1MHz to 50MHz. The dotted trace shows amplifier gain in dB. The 1dB bandwidth of this amplifier extends from 1.5MHz to well over 30MHz.

THE CIRCUIT. We will build a two stage broadband amplifier. This will be used as the low-level amplifier stages of an SSB transmitter. This amplifier is designed for an input and output impedance of 50Ω. The amplifier has a power gain of more than 30dB (1,000 times power gain). The first stage is very similar to the post-mixer amplifier stage of the 80m receiver. This stage has a gain of about 20dB. The second stage is identical to our simulated amplifier in Figure 4.12. The gain of the second stage is just over 10dB and the maximum linear output is around 400mW. Coupling between the two stages is via a 4:1 transformer.

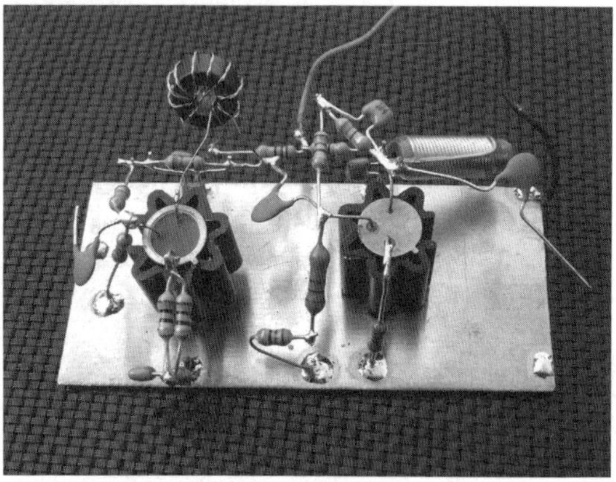

PHOTO 4.2: The layout of our amplifier.

CONSTRUCTION. The amplifier was built using dead-bug style construction on an unetched strip of PCB laminate. The layout of the amplifier can be seen in **Photo 4.2**. The schematic is shown in **Figure 4.14**. Because of the heavy use of negative feedback, the choice of transistor for Q1 and Q2 is not very critical. Any VHF/UHF transistor with hFE greater than 100, fT greater than 300MHz and Ptot of around 1-2W should work fine. UHF power transistors like the 2N5109 and 2N3866 are ideal candidates for this job. I used a pair of cheap and readily available 2N2219A transistors. The power dissipation in Q2 is quite close to the manufacturer's maximum limit. A clip-on heatsink should be used on both transistors, especially if you are using a 13-14V supply voltage. If you find that Q2 is uncomfortably hot, you can replace the 330Ω resistor in the collector-to-base feedback network with a resistor of a higher value. Note that this resistor is connected in parallel with a bypass capacitor so that it does not form part of the negative feedback signal path. T1 is a 4:1 transmission line transformer made from 10 turns of enameled copper wire, bifilar wound on an FT37-43 or similar ferrite toroid.

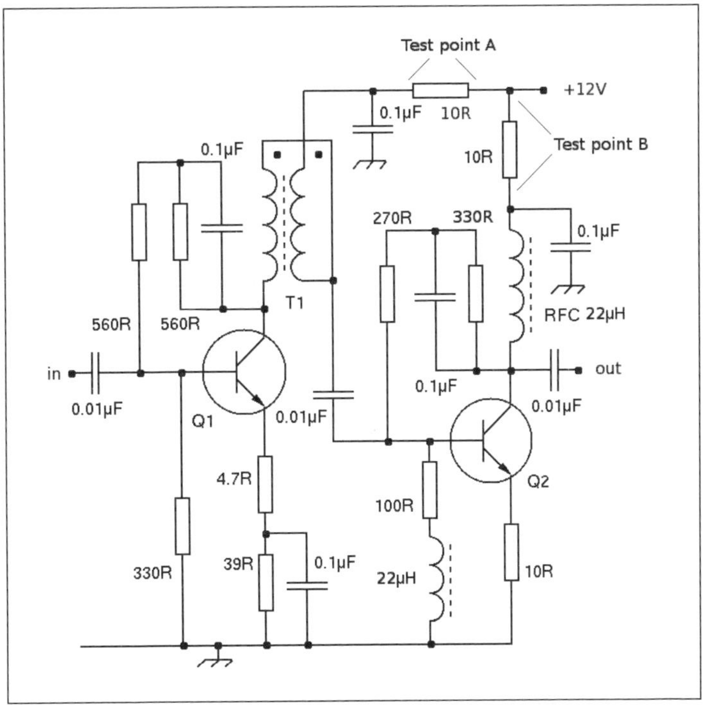

FIGURE 4.14: The schematic of our amplifier.

TESTING. The standing current in each stage can be checked by measuring the voltage drop across the 10Ω resistor in the collector supply of each of the two amplifiers. These resistors are marked as Test point A and B on the schematic. Simple application of Ohm's Law tells us that a voltage drop of 1V indicates a combined collector and base current of 100mA. With a supply voltage of 13.8V, Q1 takes about 50-60mA and Q2 takes around 80-100mA.

I used a homemade frequency standard and a precision step attenuator as a signal generator to test the performance of the amplifier. The output of the amplifier was terminated by a 47Ω resistor and the output voltage was measured with an RF voltmeter. A slightly less accurate measurement can be made using a simple diode probe and a DC voltmeter as shown in **Figure 4.15**. The tests were performed at 5, 10 and 15MHz. The test results were remarkably similar to the graph in Figure 4.13. For an input signal of 100mV RMS (-7dBm), the measured output was 3.6V RMS (+24.1dBm). This is a power gain of just over 31dB. The measured gain was identical to within a fraction of 1dB on all three frequencies.

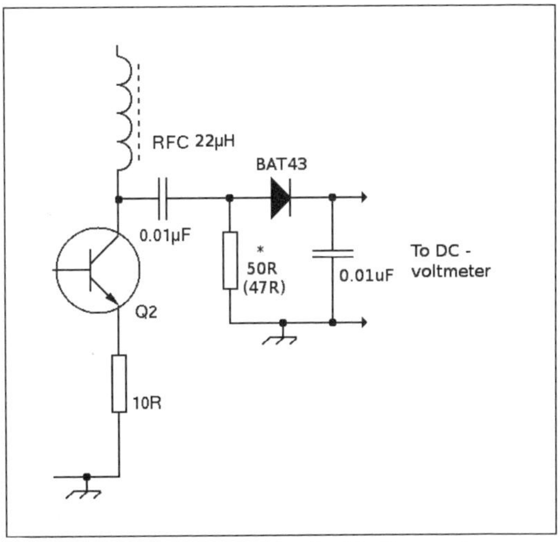

FIGURE 4.15: This simple diode probe, and a DC voltmeter, can be used to test the amplifier.

By using the step attenuator to vary the input drive level to the amplifier, I was able to find the 1dB compression point and the maximum saturated output of the amplifier. The 1dB compression point is around 400mW. If the amplifier is driven well beyond the linear region, the maximum output that can be achieved is just over 1W. Sometimes you can make surprisingly accurate measurements with little or no test equipment. Running the amplifier in its linear region produces enough

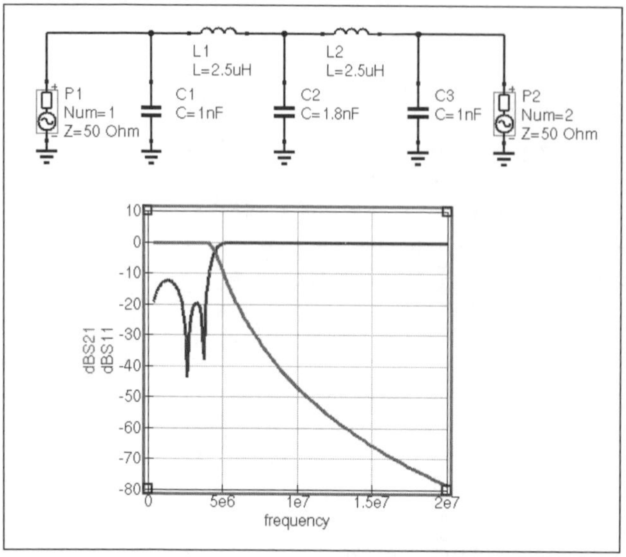

FIGURE 4.16: The low pass filter that Eamon fitted to the output of the amplifier.

power to make the 0.25W dummy load resistor quite warm. When the amplifier is run at maximum output, this resistor gets very hot.

As a final test, the 3.579MHz crystal oscillator from the SSB exciter was used to drive the amplifier input and a simple 4MHz low pass filter was fitted to the output. Details of the low pass filter are shown in **Figure 4.16**. L1 and L2 are made from 22 turns on a T50-2 powdered iron core. A Morse key was connected in series with the power supply to the oscillator to make a simple but effective CW transmitter. A single pole, single throw switch was used to switch the aerial between the transmitter and receiver. The key-down output is 1W. You can hear a sample of the CW signal at http://tinyurl.com/u92j4

Several CW contacts were made using this simple transmitter and the homebrew 80m receiver.

High Power Transistor Amplifier

POWER AMPLIFIER. Many radio amateurs consider the design and construction of high power transistor amplifiers a black art that is best left to 'the experts'. Constructors will often choose to build all of the other stages of a transmitter and use a commercially made power amplifier (PA) for the final stage. Others prefer to build the PA from a kit that includes the PCB and all other components. This is certainly a viable approach and usually results in a working PA. Buying a complete kit saves you the trouble of looking for those exotic components like special grades of ferrite, silvered-mica capacitors, RF transistors etc.

RF power transistors are usually quite expensive and are very easily destroyed by the high current and voltage levels found in RF power amplifiers. The very low input and output impedance of RF power transistors can also raise problems for the designer and constructor. I hope that you will not be discouraged by these comments. It is possible to build a good PA using readily available components. Your job will be a lot easier if you follow a few common-sense rules.

1. Take care with the thermal design of your amplifier. This usually means that you should use a big heatsink. If possible, the heatsink should be substan-

tially bigger than the required minimum. This gives you a safety margin when you are using high duty cycle modes like FM, RTTY, SSTV, PSK etc. A large heatsink also removes the need to fit a noisy cooling fan.

2. Be careful about how you interpret data-sheet information. Remember that the numbers at the top of the sheet are usually absolute maximum ratings that should never be reached in normal operation. If a transistor has a total power dissipation rating of 20W at 25°C, this does not mean that it will be capable of generating 20W of RF. Most data-sheets will have a derating curve showing the maximum allowable dissipation for a given temperature. As a rough rule of thumb, you should use a transistor with a Ptot rating of at least twice the intended RF output. Maximum collector to base voltage rating of the transistor should be several times greater than the supply voltage. The fT (unity gain, transition frequency) of the transistor should be much higher than the highest frequency that the amplifier will be used at. An fT of 5 to 10 times the signal frequency is a good figure to aim for.

3. You should also take care with the electrical design of your amplifier. The amplifier should be stable under all conditions. If your amplifier turns into a high power oscillator, it could generate destructive levels of voltage or current in the output transistor. Bipolar transistors are prone to a condition called thermal runaway, which causes current to increase with rising temperature. The increased current causes more heat leading to even greater current and so on until the transistor fails. A good PA design will have tight control of the transistor bias supply. Some designs use a thermal sensor that is mounted on the transistor header or the heatsink to reduce the bias when the PA transistor gets hot.

4. Because of the very low impedances involved, the physical layout of the circuit is very important. Small amounts of stray resistance or reactance that would have little or no effect on a low-level amplifier can have a significant impact on the performance of a power amplifier. Good de-coupling of the power supply is difficult to achieve when DC and RF currents of several amps are flowing in the circuit. Thick conductors or wide PCB tracks should be used to keep stray resistance and inductance to a minimum. Ground connections should be short and straight. Good VHF/UHF practice should be followed, even when you are building an amplifier for the lower frequency bands. The physical layout of push-pull amplifiers should be as symmetrical as possible to ensure that the circuit is electrically and thermally balanced.

5. Take great care when testing and adjusting the amplifier. RF power amplifiers are usually quite robust once they have been properly adjusted and the lid is screwed on the box. Most of the semiconductors that I have destroyed in the past have failed during initial testing and adjustment. Be careful when you are setting the bias voltage for the correct standing current. Watch out for any signs of thermal runaway or parasitic oscillation. Take care when you are tweaking the amplifier to try and get just a little bit more output. This is by far the most dangerous time in the life of an RF PA.

POWER AMPLIFIER CIRCUITS. Most RF power amplifiers use one or more transistors in the common-emitter configuration (common-source for

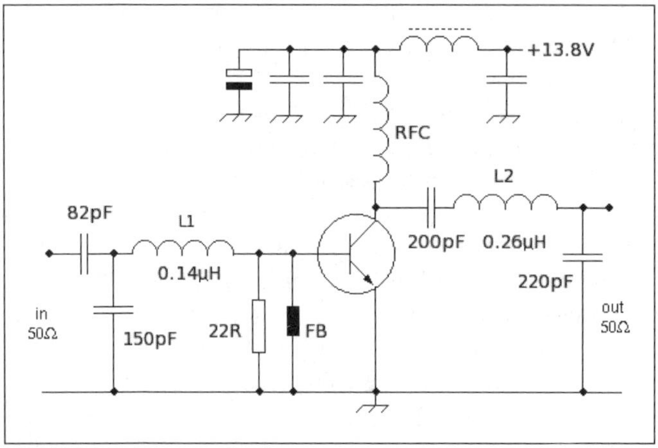

FIGURE 4.17: a typical arrangement for a single ended, Class C power amplifier.

FETs). Single ended designs can use a single transistor or several transistors connected in parallel. Push-pull amplifiers will usually use two devices. Some high power designs will combine these two methods by using two groups of parallel connected transistors in a push-pull arrangement. **Figure 4.17** shows a typical arrangement for a single ended, Class C power amplifier. The transistor is an NPN bipolar junction transistor (BJT) connected in the common-emitter configuration. The input impedance of RF power transistors is quite low. Typical values range from a few ohms at medium power levels to less than 1Ω for high power devices. For amplifiers with a 12V supply, the output load impedance will be in the same range. The optimum load impedance is approximated by $Vcc^2/2Po$ where Vcc^2 is the supply voltage squared and 2Po is twice the output power of the amplifier. If the amplifier in Figure 4.17 was designed for an output of 10W at a supply voltage of 12V, the output load impedance should be 7.2Ω . A 100W amplifier of similar design would have an output load impedance of 0.72Ω! The input impedance will be determined by the output power level and the gain of the transistor at the signal frequency. Transistor current gain decreases at higher frequencies. This means that at MF/HF frequencies, the transistor input impedance gets lower as frequency is increased. This is not necessarily the case at VHF/UHF where the input capacitance of the transistor and the inductance of the base connector lead have a more significant effect on the input impedance.

We will assume that our hypothetical amplifier in Figure 4.17 uses a 2SC1969 transistor that has a maximum power dissipation rating of 20W. The manufacturer's data-sheet does not specify fT, but it claims that the device should have a power gain of >12dB at 27MHz. The data-sheet specifies hFE as 10-180 with a typical value of 50. If this amplifier is powered from a 13.8V supply and has a maximum output of 10W at a frequency of 28.5MHz, the optimum load impedance will be around 9.5Ω. The input impedance of the transistor is something of a moving target. It will vary quite dramatically with drive level, frequency and temperature. The peak input voltage of a bipolar transistor is quite low, usually less than 1V. fT/f gives a good approximation of the high frequency current gain of a transistor. We will assume that our 2SC1969 has a current gain of about 10 at 28.5MHz. We will make the rather pessimistic assumption that the amplifier will have an efficiency of about 50%. This means that the DC input power to the amplifier will be 20W and the collector current will be about 1.5A. The base current will be around

150mA and the input resistance will be around 5 ohms. This is a typical value for a 10W PA. Like most RF power transistors, the 2SC1969 is emitter ballasted. This means that there is a built-in resistor connected in series with the emitter (or emitters) of the device. The degenerative feedback caused by this resistor will improve the linearity and reliability of the transistor. Many RF power transistors are constructed from arrays of smaller transistors on a single die. The emitter ballast resistance encourages equal current sharing between the emitters. Note that the input and output impedances are not a pure resistance. The input and output capacitance of a medium power transistor is usually in the region of 50-150pF. The inductance of the base lead can also have a significant effect, especially at VHF and above. The design of input/output matching networks should make allowance for these reactances. At MF/HF frequencies, the effect of these reactances is often small enough that they can simply be ignored.

MATCHING NETWORKS. Radio transmitters are usually made up from several separate stages. This is generally true even when the entire transmitter is constructed on a single PCB. Inter-stage connections are usually made at an impedance of 50Ω. This simplifies the design and testing of each individual stage. Using a standard I/O impedance also simplifies the task of designing and testing band pass and low pass filters. Broadband amplifiers often use transformers with a ferrite core for I/O matching. Transmission line transformers like the ones used in some of our recent Homebrew projects are suitable for use in broadband PAs. Conventional transformers with a separate primary and secondary winding can also be used. The use of a high permeability ferrite core allows a transformer winding with very few turns to have enough inductance for use at lower frequencies. Some broadband transformer designs only use a single turn for the low impedance winding. This results in a coil that has very low self capacitance and therefore has a very high self resonant frequency. Such transformers can be used over a wide frequency range.

Narrow band or single band designs like the example in Figure 4.17 are designed to operate over a limited range of frequencies. The relatively low input and output impedances can be transformed to 50Ω by using LC (inductance/capacitance) matching networks. There is a bewildering array of different networks that can be used for this job. The well known PI network - often used in valve PAs - can also be used for matching transistor circuits. Because of the very low impedances encountered in high power transistor amplifiers, the PI network often requires unreasonably large or small values of L or C. Other networks that are derived from the L or T network are more commonly used for I/O matching. The simplest arrangement would use one inductor and one capacitor configured as an L-network. The low pass type of L-network with a series L and shunt C is sometimes used because it gives some harmonic rejection. High pass L-networks with series C and shunt L can also be used. Simple L-networks are rarely used in high power PAs because they give little or no control of the loaded Q of the network. Adding a third reactive element allows the designer to have some control of the loaded

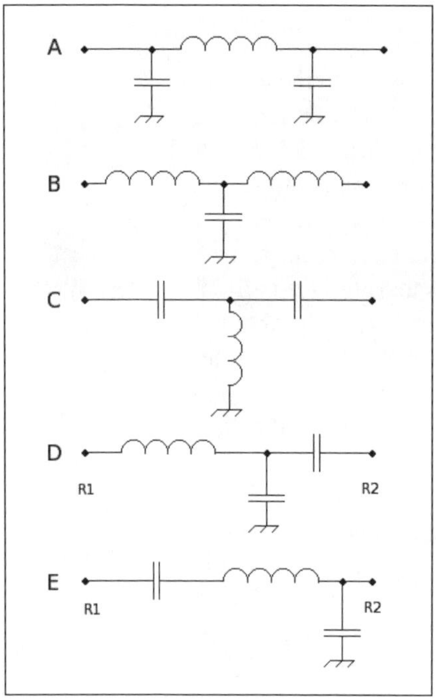

FIGURE 4.18: A selection of the most commonly used matching networks.

Q. **Figure 4.18** shows a few of the most commonly used matching networks. Figure 4.18A is a PI network, b is a low pass T-network, c is a high pass T-network, d and e are controlled Q networks that are derived from the L-network. Such networks are often named according to the order in which the reactive elements are arranged. Figure 4.18D is known as the LCC network or CCL network, depending on which side you are looking at. Note that network D and E can only be used if R2 is greater than R1.

We will use a CCL network to match the 50Ω input impedance of our power amplifier to the 5Ω input impedance of the transistor. I used the formula from chapter 5 of SSD **[40]** to calculate the component values. Similar formulae can be found in the *RSGB VHF/UHF manual* and the *ARRL Handbook*. The network from Figure 4.18E will be used to match the 9.5Ω collector load impedance to 50Ω. The component values are shown in Figure 4.17.

You should avoid using excessively high values of loaded Q for transistor matching networks. High-Q networks will greatly increase the risk of instability and will also lead to increased losses. Transistor I/O networks should only use a loaded Q of slightly more than $\sqrt{((R1/R2)-1)}$ which is the minimum Q that is needed to achieve the required impedance transformation. Typical Q values for transistor power amplifier matching networks range from 3 to 10. The input and output networks for our amplifier were both designed for a loaded Q of 5. **Figure 4.19** shows a plot of the frequency response of both networks. The dotted trace is for the input network and the dashed trace is for the output network.

The transistor bias supply for low power amplifiers is usually provided by a simple resistor network. This approach is not suitable for use in power amplifiers because the relatively high base current of a typical PA would lead to excessive voltage variations in the base supply. It is common practice to provide the bias supply via a small inductor. This inductor should have XL (inductive reactance) several times greater than the input impedance of the transistor. A few turns of wire on a small multi-hole ferrite bead or toroid is ideal for this purpose. Since our 10W amplifier is designed to operate in Class C, this inductor connects the base of the

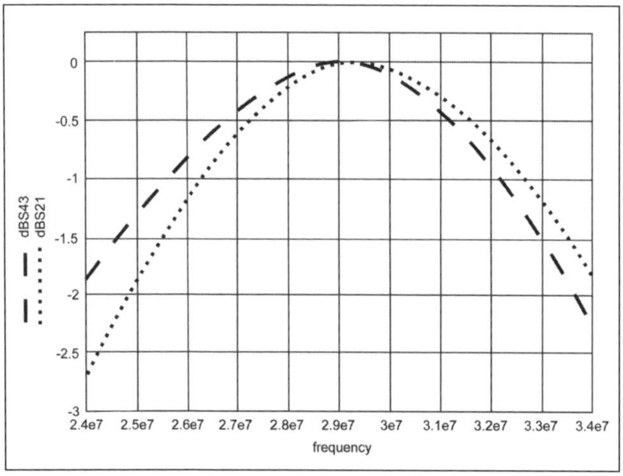

FIGURE 4.19: A plot of the frequency response of the input and output networks of our amplifier.

transistor to ground. The 22Ω resistor connected in parallel with this inductor helps to suppress parasitic oscillations caused by unwanted resonances in the base circuit. This resistor also helps to stabilise the input impedance of the transistor which tends to vary with drive level and frequency. The price we have to pay for this improved stability is a slight loss of gain which is due to the power dissipated in this resistor. Note that the DC side of the collector load inductor (RFC) is bypassed by several capacitors connected in parallel. This is to ensure adequate decoupling of the DC supply at all frequencies.

PHOTO 4.3: The layout of our amplifier.

BROADBAND MF/HF PA. Now we will leave the virtual world of pocket calculators and simulation software behind. This project is an HF linear amplifier with a maximum output of 20W. The amplifier provides more than 20dB of gain from 1.8MHz to 30MHz. The driver stage uses a 2SC2166 transistor operating in Class A. The Class AB final stage uses a pair of 2SC1969 transistors in a push-pull configuration. Ferrite cored broadband transformers are used for I/O and inter-stage matching. The schematic of the amplifier is shown in **Figure 4.20** and the layout can be seen in **Photo 4.3**.

CONSTRUCTION. The amplifier was built dead-bug style on a strip of unetched PCB laminate. Rectangular holes were cut in the PCB so that the transistors could be mounted directly on the heatsink. The prototype was built on the back of a large

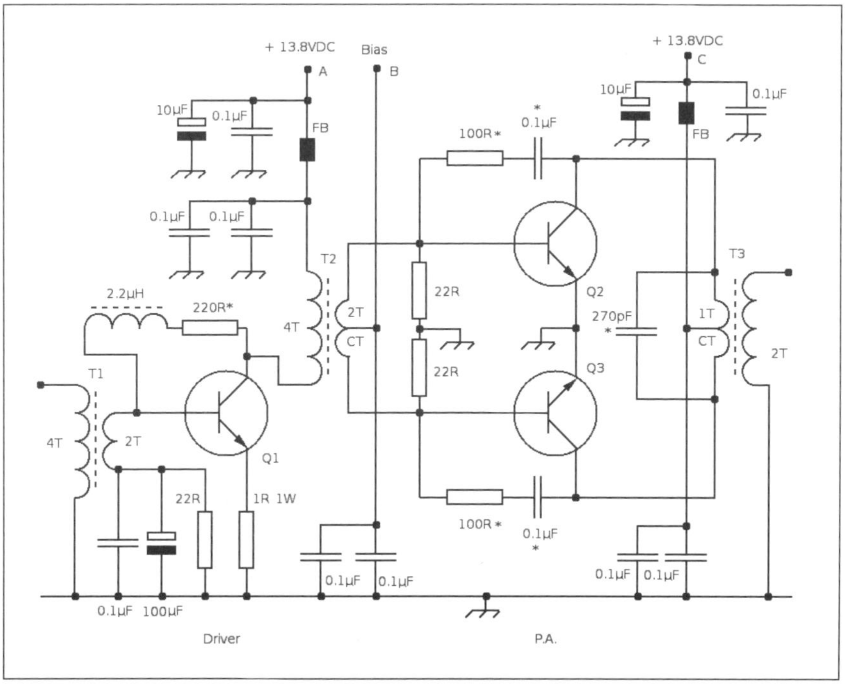

FIGURE 4.20: The schematic of the amplifier.

PHOTO 4.4: The mounted transistors.

PHOTO 4.5: The core of T3 is a pair of Maplin N91AB ferrite sleeves.

0.8°/W heatsink. The final version uses a 1.8°/W heatsink (Maplin FL42V or similar). The amplifier is enclosed in an aluminium box (Maplin AB13 or similar).

A large rectangular slot was cut in the bottom of the box to facilitate mounting of the transistors. Each transistor is electrically isolated from the heatsink by an insulated mounting kit. Mica or silicone-rubber insulators can be used. A very thin film of heatsink compound should be used on both sides of mica insulators.

Use a sharp 3.5mm drill for drilling the holes in the heatsink. Always start drilling on the transistor side of the heatsink. You can use a large drill-bit (8-10mm) as a hand tool for removing any rough edges around the 3.5mm transistor mounting holes. Make sure that the heatsink is absolutely smooth, flat and clean before you mount the transistors. **Photo 4.4** shows the mounted transistors.

The broadband transformers are wound on high permeability ferrite cores. T1 has 4 turns of insulated wire for the primary and two turns for the secondary. The core for T1 is made from 4 T37-43 toroids. T2 is also wound on 4 T37-43 toroid cores. The primary of T2 is 4 turns, the secondary is 2 turns, centre tapped. The primary of T3 uses a single turn of 2.5mm insulated wire. I used standard green/yellow earth cable for the centre tapped primary. The secondary is 2 turns of 1mm insulated wire. The core of T3 is a pair of Maplin N91AB ferrite sleeves. **Photo 4.5** shows T3. The 0.1µF capacitors must be good quality ceramic types with a maximum voltage rating of at least 100V. The two capacitors in the shunt feedback network of the PA transistors get quite warm during normal operation. A short-circuit failure of one of these capacitors could cause a potentially destructive increase in base current! The 270pF capacitor is a silvered mica or other low loss type. The 220Ω and 100Ω resistors are 2W metal film or carbon film. All other resistors are carbon film or other low inductance types with a power rating of 1W. The 2SC1969 transistors should be well matched for current gain. Try to get a pair from the same batch with the same hFE group markings. Matched pairs are often available on ebay. The 2SC1969 is available from Grandata, Sycom, Donberg and several other suppliers. Other CB PA transistors like the 2SC1307 and 2SC2312 would probably work well in this circuit. The inductor in the shunt feedback network of the driver transistor is a 2.2µF RF choke (Maplin WH31J).

For Class AB operation, the amplifier needs a bias supply of about 0.7V. A schematic of the bias supply is shown in **Figure 4.21**. This is an active supply that uses a BD135 transistor to provide base current to the PA transistors. The BC547 transistor limits the output voltage to about 0.7V. A 1kΩ pot is used to set the standing current in the PA transistors.

TESTING AND ALIGNMENT. The bias circuit should be built and tested separately before it is connected to the PA transistors. It should produce an output voltage of around 700mV. The 1kΩ pot should be set for maximum resistance. Check the circuit carefully before you connect it to the PA.

Unless you have a power supply with adjustable current limiting, it is a good idea to place a small resistor in the positive lead from the power supply. A 1.8 or 2.2Ω resistor will limit the maximum current to a few amps. This might be enough to save the transistors from destruction in the event of an accident.

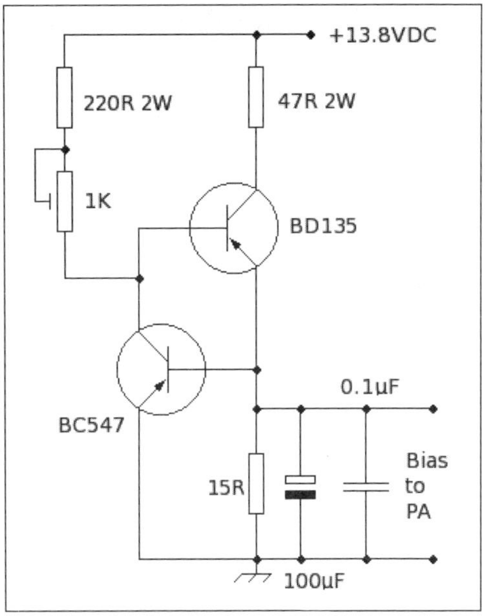

FIGURE 4.21: A schematic of the bias supply for our amplifier.

To check the DC bias of the driver stage, you should connect the 13.8V supply to the driver stage only. This is point A on the schematic. The transistor should run warm but not very hot. Check the voltage across the emitter resistor to confirm that the emitter current is around 0.5A. A wet fingertip makes a very sensitive probe for checking the temperature of the transistor package. If your finger gets burned, you can regard it as punishment for not mounting the transistor properly!

Once all is well with the driver stage, you should connect the bias supply to point B and the 13.8V supply to point C. Proceed with care, if you were to accidentally apply 13.8V to the bias supply, you would almost certainly destroy the PA transistors. The total current consumption of the PA module should be less than 1 amp. You should now adjust the 1k pot in the bias circuit for a total current of 1A: 500mA for the driver and 250mA each for the two PA transistors. Keep an eye on the current meter while you are adjusting the PA standing current. Leave the amplifier powered up for a few minutes and check the temperature of the PA transistor packages. Both should be warm but not hot.

Once the DC bias conditions have been established, the PA is ready for use. I used the low level output from my home made HF transceiver as a signal generator for testing the PA. A 50Ω dummy load and a diode probe were used for power measurement. My makeshift signal generator produces an output of 100-120mW. At this drive level, the measured output of the amplifier was more than 20W from 1.8MHz to 25MHz. The gain starts to fall off slightly at higher frequencies. Maximum output at 29MHz was 14W. I used a 7-pole LPF between the output of the PA and the aerial for on-air testing on 80m. Reports on the signal quality have been very good. Using a homebrew 80m receiver as a poor man's spectrum analyser confirms that there are no audible IMD products at more than ±3kHz. The Mitsubishi 2SC1969 data-sheet says that the 2SC1969 can withstand an infinite load VSWR. The amplifier has survived several tests at maximum output into an open or short-circuit load.

13.8V DC Power Supply

This power supply features excellent voltage regulation, current limiting, over-voltage protection, low noise output and good immunity to RF. Before we start soldering, we will take a closer look at some of the basic principles involved.

VOLTAGE REGULATORS. An ideal diode would have a very low voltage drop when forward biased and a very high breakdown voltage when reverse biased. Modern rectifier diodes come close to this ideal. High voltage rectifier diodes like the 1N4007 have a reverse breakdown voltage of more than 1,000V. The Zener diode is a special type of diode that is deliberately designed to have a low reverse breakdown voltage. Most Zener diodes are heavily doped silicon junction diodes. The reverse breakdown voltage is dependant on the type and amount of impurities added to the P and N silicon semiconductor during the manufacturing process. **Figure 4.22** shows a plot of current vs voltage for a 5.6V Zener diode. When the diode is forward biased, it behaves in much the same way as an ordinary silicon rectifier or switching diode. There is very little current flow until the forward voltage reaches about 600mV. As the voltage is increased beyond this point, the diode current increases rapidly. When the diode is reversed biased, there is very little current flowing through the diode until the reverse breakdown voltage is reached. The transition from low to high current occurs at a very sharply defined voltage. This voltage is known as the Zener knee or Zener voltage. Because the reverse breakdown voltage of a Zener diode is relatively stable over a wide range of diode current, this type of diode makes an excellent voltage reference.

Another device that has similar characteristics is the avalanche diode. The distinction between Zener and avalanche diodes has become rather blurred because all reverse biased, voltage reference diodes are usually referred to as Zeners, regardless of whether the avalanche or Zener effect is the dominant mode of operation. Avalanche breakdown occurs in semiconductors or even in good insulators when the voltage across the material is great enough to accelerate free electrons to a level which causes them to strike other atoms and dislodge more free electrons. This leads to an avalanche effect which turns a semiconductor or an insulator into a good conductor.

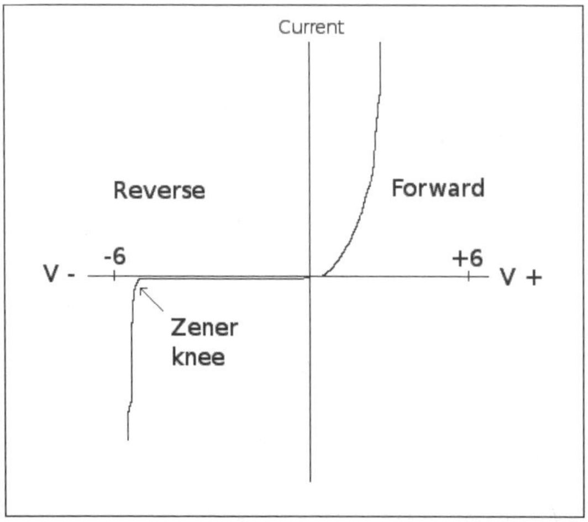

FIGURE 4.22: A plot of current vs voltage for a 5.6V Zener diode.

Zener diodes are low voltage devices and avalanche diodes are higher voltage devices. Both Zener and avalanche effects can be found in all junction diodes. Zener breakdown is the dominant mode in diodes with a reverse breakdown voltage of less than 5.6V. Avalanche breakdown is the dominant mode in diodes with a reverse breakdown voltage of more than

5.6V. Because Zener diodes have a negative temperature coefficient and avalanche diodes have a positive temperature coefficient, Diodes with a Zener knee of around 5.6V have very good voltage stability over a wide temperature range. This is why 5.6V and 6.8V have become the standard Zener voltages for temperature sensitive applications.

Figure 4.23 shows the classic Zener diode voltage regulator circuit. Resistor R limits the current through the Zener diode to a safe value. Capacitor C acts as a smoothing capacitor to improve the dynamic regulation. This capacitor also helps to filter out noise generated in the Zener diode. This circuit will provide good voltage regulation provided that the current through the Zener diode is large compared with the current through the output load. At high load currents, the voltage drop across R will increase and the current through the Zener diode will decrease. This will eventually lead to a loss of regulation. Since the voltage across the Zener diode is the same as the voltage across the output load and the current through the Zener diode is always greater than the current through the output load, this type of regulator is very inefficient. Most of the power from the unregulated supply is dissipated as heat in the Zener diode and the current limiting resistor. This is not a serious problem in low-power circuits where the load current is small.

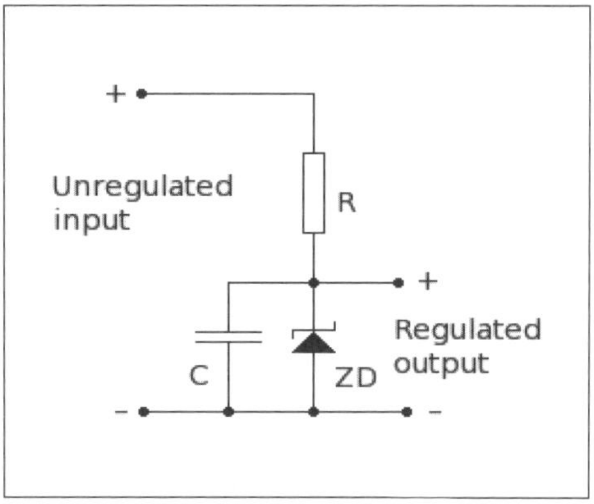

FIGURE 4.23: The classic Zener diode voltage regulator circuit.

It is possible to provide higher load currents by using a high power Zener diode and reducing the value of R. This approach is usually impractical because high power Zeners are difficult to find and quite expensive. The most common type of Zener diode has a maximum power dissipation of 1.2W. It is possible to use a low power Zener in combination with a power transistor to make a high power shunt regulator. This 'Super Zener' circuit is shown in **Figure 4.24**. The output voltage of this type of regulator is the Zener voltage plus the transistor's base to emitter voltage drop of about 700mV. The current through the Zener diode is slightly more than the transistor emitter current divided by the transistor beta. This circuit is often used to provide cathode bias in high power valve amplifiers.

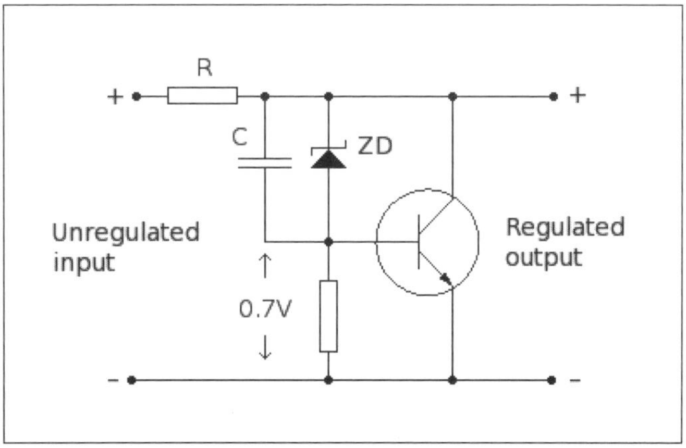

FIGURE 4.24: The Super Zener circuit.

The shunt regulators described above are all relatively inefficient. Using a series transistor as the regulating element will lead to a big improvement in efficiency. **Figure 4.25** shows a simple regulator using a series transistor to control the output voltage. Power dissipation in the transistor is proportional to the output current. This is the opposite to the shunt regulator which has maximum dissipation in the shunt transistor or diode when the output current is zero. The output voltage of the regulator in Figure 4.25 is the Zener voltage minus the emitter to base voltage drop of the transistor. Since a typical power transistor has a DC current gain of 50-200, this circuit can deliver substantial output current. We will consider a simple design based on this circuit.

We will assume that the unregulated voltage is 25VDC and the required output voltage is around 13 to 14VDC. We will use a 14V Zener diode with a maximum power dissipation of 1.2W. Running the Zener diode at close to its maximum power rating would result in poor voltage stability because of the high temperatures generated at the diode junction. We will choose a value of R that limits the diode current to about 20mA. The voltage across R will be the unregulated voltage of 25V minus the Zener voltage of 14V (25-14)/0.020 gives us a value of 550Ω . We will use the nearest standard value which is 560Ω. Power dissipation in the resistor is 216mW so a 0.5W resistor will be adequate for the job. To achieve reasonably good voltage regulation, we will limit the base current of the pass transistor to a maximum of 10mA. If the transistor has hFE of 100 which is a typical value for a medium power transistor, this simple regulator will be able to deliver about 1A to the output load. The output voltage will be 14V minus 0.7V = 13.3VDC. This is not an ideal regulator because the E-B voltage drop of the transistor will tend to vary with the output current. Voltage regulation can be improved slightly by connecting a resistor across the output so that some current is always flowing in the transistor, even when no current is drawn by the ex-

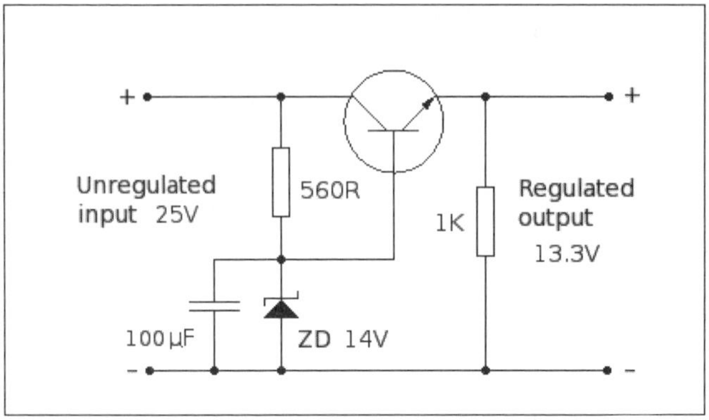

FIGURE 4.25: A simple regulator using a series transistor to control the output voltage.

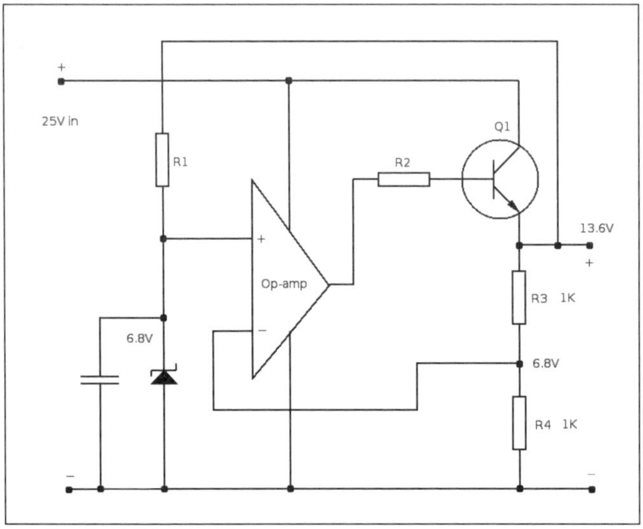

FIGURE 4.26: A voltage regulator that uses a Zener diode as a voltage reference and an op-amp (operational amplifier) as the error amplifier.

ternal load. Suggested component values are shown in the schematic. Another problem with this simple design is the fact that the performance of the regulator is dependant on the gain of the transistor. Transistor hFE can vary quite dramatically, even when transistors are of the same type and brand.

High performance voltage regulators usually use a precise voltage source like a Zener diode operated under ideal conditions as a voltage reference. The output voltage of the regulated supply is compared to this precise reference and any errors are compensated for by an error amplifier. The circuit in **Figure 4.26** shows a voltage regulator which uses a Zener diode as a voltage reference and an op-amp (operational amplifier) as the error amplifier. Once again, the off-load unregulated input is 25VDC. The voltage reference is a 6.8V Zener diode. In order to achieve the best voltage stability, the Zener should be fed from a constant-current source. This is achieved by taking the Zener current from the stabilised output of the regulator. As the Zener voltage is exactly 6.8V and the output voltage is exactly twice this value (13.6V), the voltage across R1 will be 6.8V and the current through R1 will remain constant, regardless of the output load current. Or, to put it another way, the current through the Zener will be defined by its own reverse breakdown voltage. R1 should be a low temperature coefficent type. Modern 1% tolerance, metal-film resistors are very suitable for this application. R3 and R4 form a voltage divider which reduces the output voltage to 6.8V. Errors caused by temperature related changes in the resistance of R3 and R4 will be cancelled out, provided that both resistors are identical and both resistors are at the same temperature. Since the input current of modern MOSFET input op-amps is measured in picoamperes, there is virtually no current drawn from the voltage reference. This means that the reference voltage will be almost completely independent of the output current and the output voltage can be very accurately controlled. Even the few mV of input offset voltage of a modern op-amp can be adjusted out by using an offset null circuit. The maximum output current can be increased by using a Darlington transistor for Q1 or by using a pair of transistors in a Darlington configuration. Note that the voltage drop across R2 and the E-B voltage drop of Q1 are automatically compensated for by the error amplifier. I have found that the day to day voltage stability of simple Zener diode regulators is often better than 1%. If you really do need better stability, the Zener diode can be replaced by a precision voltage reference IC. State-of-the-art devices have a temperature coefficent of a few ppm/degC. For the ultimate in voltage stability, the voltage reference can be mounted in a temperature controlled oven.

IC Regulators

The type of regulator described previously is typical of the circuits used in IC (integrated circuit) voltage regulators. The voltage reference, error amp and pass transistors are all combined into a single package. Most IC voltage regulators also include current limiting and/or thermal shutdown circuitry. The well known 78xx series of 1A regulators has all of these features combined into a three-pin package. More flexible devices like the LM723 include features like variable output voltage and variable current limiting.

A Homebrew 13.8V Power Supply

This is our first mains-powered Homebrew project. Please take care when working with mains voltages. A shock from the 230V mains can cause serious injury or even death. Make sure you use components which are rated for mains voltage and that all live components are well insulated. Make sure that the metal chassis of this project is properly grounded and that both the mains and the low voltage side of the transformer are protected by a suitably rated fuse. Always disconnect equipment from the mains before you work on it. Do not wear metal jewellery when working on high current supplies.

The first and most import consideration when designing and building a power supply is safety. The power supply should be properly insulated from the live mains. All live components must be safely covered and the chassis and case of the supply must be properly grounded to the mains earth. Components must be rated for mains use. Low voltage components can cause a fire if they are exposed to mains voltages. The equipment should use suitably rated fuses and as a last line of defence, the 13A mains plug should be fitted with an appropriate fuse. A 13A fuse is unlikely to give much protection to a relatively low power circuit. Low voltage circuitry should be well isolated from the mains earth. Don't take unnecessary risks like having mains voltages on the same PCB as the low voltage regulator circuit.

The second consideration is the protection of equipment which is connected to the DC power supply. If excessive current is drawn from the supply, it should limit the output voltage or preferably shut down altogether. Even if the power supply is fitted with current limiting or electronic trip circuitry, a suitable fuse should be fitted as a backup. You should always have some form of over-voltage protection.

The third and least important consideration is the protection of the power supply itself. The job of the power supply is to provide a clean and stable supply of the correct voltage to the equipment it provides power for. All electronic components will fail eventually. Hopefully your homebrew power supply will give many long years of safe and reliable service. However, if a component should fail, you need to make sure that the power supply fails in a safe mode. In the event of a regulator failure, a crowbar over voltage protection circuit will short-circuit the output of the supply and prevent unregulated voltage from reaching your valuable equipment. The high current through the crowbar will operate the current limiting or blow the fuse. Even in the event of a major failure, power supplies should always fail in a 'heroic' mode. This means that the supply will damage itself before it will damage external equipment.

THE CIRCUIT. **Figure 4.27** shows the mains transformer and rectifier circuit. The mains input is via an IEC (kettle type) connector. IEC connectors with an integrated RFI/EMI filter are preferable to plain non-filtered types. The earth connection of the IEC connector should

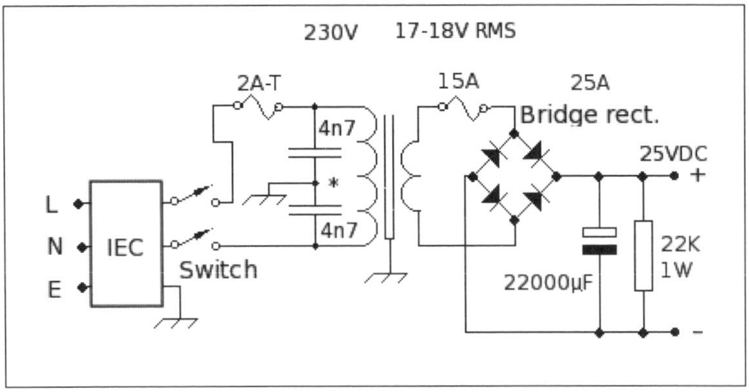

FIGURE 4.27: The mains transformer and rectifier circuit for our homebrew project.

be well bonded to the metal case of the power supply. The mains switch should be a double-pole type rated for at least 3A at 250V. The live line should be fitted with a suitable fuse. I used a 2A - T (time delay) fuse. A quick-blow fuse would offer greater protection, but it might not be able to cope with the large inrush current when the unit is first powered up. Some IEC connectors like the Maplin FT37S have a built-in 20mm fuse holder. The mains transformer should have a primary voltage that matches your local mains supply. The European standard is 230V RMS. My local mains supply is usually closer to 240V RMS. The transformer should have a secondary voltage of 17 to 18V RMS (25V peak). The 4700pF capacitors in the primary circuit of the transformer should be Y2 types that are properly rated for line-to-ground applications. Don't be tempted to use ordinary capacitors from the junk box. These capacitors should be omitted if an IEC connector with a built in RFI filter is used.

The connections to the IEC connector, switch, fuse holder and transformer should all be insulated with heat-shrink sleeving. Note that the chassis mounted IEC connector is called the plug and the line mounted part is the socket!

The size of the transformer will depend on the maximum current that the supply will be expected to deliver. I used a 250VA transformer with a secondary voltage of 18V RMS. This transformer will maintain good regulation at output current levels of up to 14A. I used a 22,000µF capacitor with a voltage rating of 40VDC. If large electrolytic capacitors are not available, you can use two or more smaller capacitors connected in parallel. As the tolerance range of electrolytic capacitors is usually quite large, the exact value of the smoothing capacitor is not very critical. Somewhere between 20,000µF and 30,000µF will be suitable for a 10A supply. The off-load voltage of the unregulated DC supply will be 1.414 times the RMS voltage from the transformer, minus the 1.4 volt drop of the diodes in the bridge rectifier. I get just over 24V from mine.

Now I must mention a very important point. The negative output from the unregulated 25VDC supply must NOT be connected to ground.

VOLTAGE REGULATOR. A CA3140 MOS input op-amp is used as the regulator/error-amplifier. The operation of this circuit is much the same

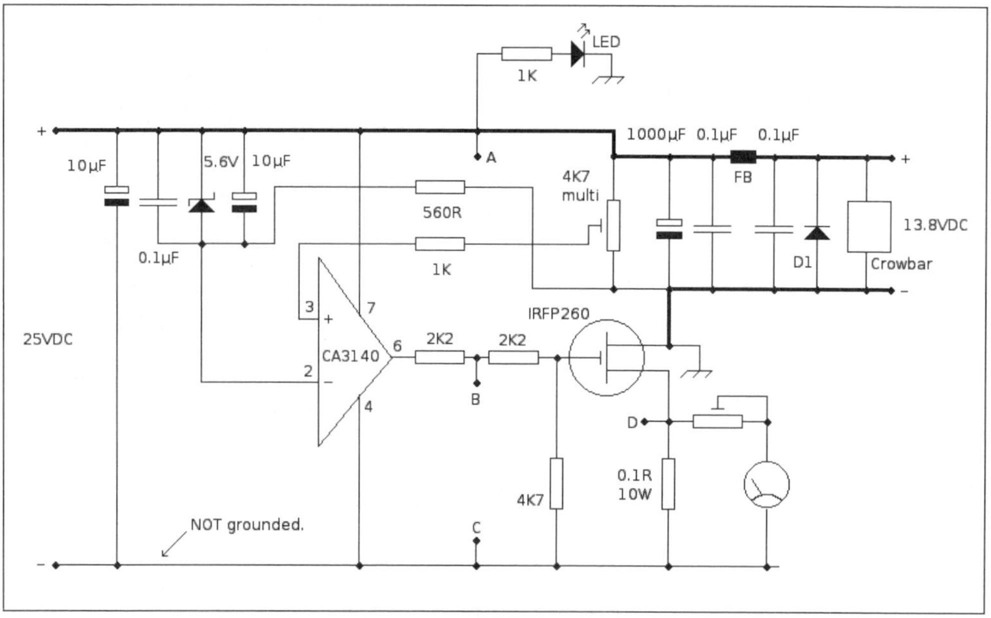

FIGURE 4.28: A schematic of the voltage regulator.

as the op-amp regulator described earlier. **Figure 4.28** shows the schematic of the voltage regulator. A power MOSFET is used as the series pass transistor instead of the bipolar transistors used in earlier examples. The MOSFET's ability to withstand huge current surges makes it an ideal device for this application. It is common practice to put the pass transistor on the positive side of the DC output. I decided to turn the entire circuit upside-down by placing the pass transistor on the negative side. This arrangement offers a few advantages. The gate to source voltage of the MOSFET is around 4 to 5V when the output is under load. This means that the regulator would need a voltage overhead of greater than this value to maintain regulation at maximum load. This requirement is eliminated by putting the MOSFET in the negative rail. Another big advantage with this arrangement is that the MOSFET can be mounted directly on the heatsink without any insulating hardware. Note that the negative rail on the unregulated side of the supply operates at about 10V below ground potential so it must not be connected to ground. ABC and D are the connection points for the optional over-current trip (electronic fuse) circuit.

CONSTRUCTION. The power supply should be built in a metal enclosure. The earth terminal of the IEC connector and the metal core of the mains transformer should be well bonded to the case/chassis. A thin film of heatsink compound should be applied to the bottom of the bridge rectifier before it is bolted to the chassis or heatsink.

I built the regulator circuit on a small strip of Veroboard. The IRFP260 MOSFET should be mounted on a very substantial heatsink. A thin film of heatsink compound should be used between the transistor and heatsink. I used a flat washer between the M3 mounting screw and the top of the

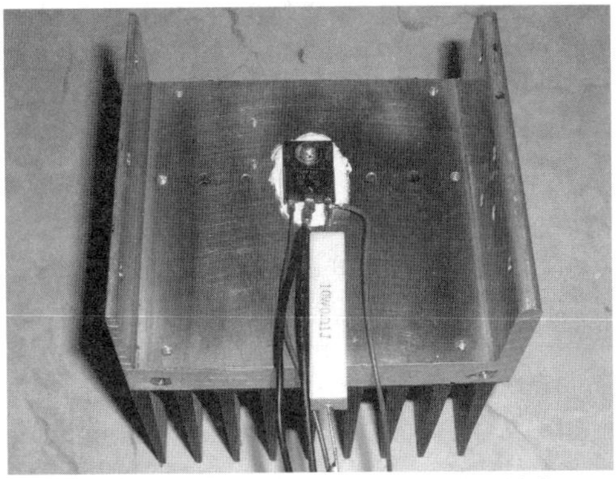

PHOTO 4.6: The mounted MOSFET for our power supply.

transistor. **Photo 4.6** shows the mounted MOSFET. The 15A fuse is a plug-in type as used for car electrics.

The positive supply rail is common to the unregulated input and the regulated output. The regulated output voltage and the Zener diode reference are both referenced to the positive rail. You should use fairly thick wire for this connection. Ideally, all connections to the positive rail should be made at a common point close to the DC output terminals. Note that the positive rail is represented by a thick line in the schematic. The component marked FB in the schematic is a ferrite toroid or ferrite sleeve. D1 is a rectifier diode which helps to protect against back-EMF from inductive loads or accidental connection to a voltage source of the opposite polarity. I used a 1N5408 for this purpose.

TESTING. You should test the transformer, rectifier and smoothing capacitor section before it is connected to the rest of the circuit. After checking all connections carefully, you should connect the mains supply and switch it on. Check that the DC voltage across the smoothing capacitor is approximately 25VDC. Leave the unit powered up for a few minutes to make sure nothing is getting hot. If all seems well, switch off the unit and disconnect it from the mains. Connect the 25V supply to the regulator and pass transistor and then power the unit up again. The power LED should now be illuminated. Adjust the 4.7k multi-turn pot until the output voltage is exactly 13.8VDC.

I tested the voltage regulation of the supply using a homebrew dummy load. Another IRFP260 was used as the load. I was planning to draw a nice graph of output voltage vs output current. I measured the voltage across the output terminals at 1A intervals from 0 to 12A. The voltage at all current levels from 0 to 11A was exactly 13.80VDC. The over-current trip activated at 11.75A. The voltage regulation is better than my ability to measure it. My digital multimeter is limited to a resolution of just 10mV. This performance was far better than I expected and quite a bit better than my existing 13.8V supply. Since my proposed V/I graph would consist of a perfectly straight line, I have abandoned my graph drawing plans for now.

Over-current trip (electronic fuse) and Over-voltage Protection (crowbar) Circuits for the DC Power Supply

OVER-CURRENT PROTECTION. Both of the protection circuits are based on the SCR (Silicon Controlled Rectifier). The SCR (also known as the thyristor) is a semiconductor device which can be used as a high speed switch.

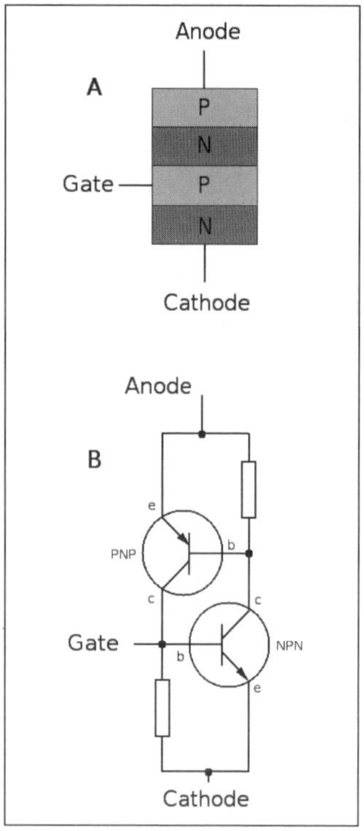

FIGURE 4.29: Physical construction and theoretical circuit of a thyristor (SCR).

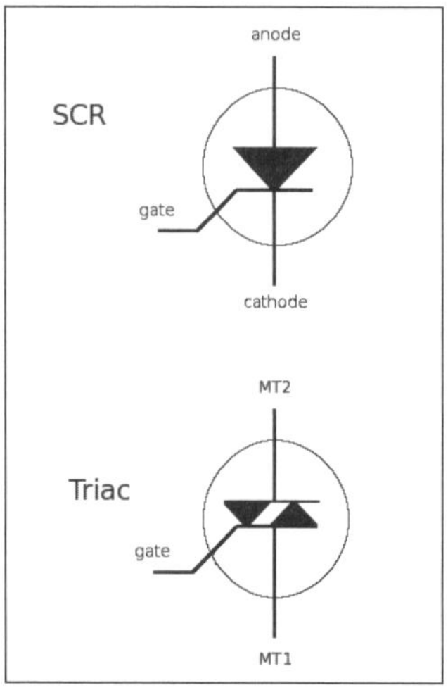

FIGURE 4.30: Circuit symbols for thyristor (SCR) and triac.

Figure 4.29A shows the internal construction of a SCR. The device is made from four layers of P and N type silicon or other semiconducting material. The SCR behaves like a pair of cross-connected transistors as shown in **Figure 4.29B**. The SCR will conduct when a trigger voltage is applied to the gate terminal. The device will remain in this switched-on state, even after the gate trigger voltage is removed.

Consider what would happen if you were to connect a DC supply across the circuit in Figure 4.29b. The positive side of the DC supply would be connected to the anode terminal and the negative side of the DC supply would be connected to the cathode terminal. Since both transistors are biased off by resistors connected between the base and emitter of each transistor, little or no current will flow through the circuit. If a positive pulse with a peak voltage of more than about 700mV is applied to the gate terminal, the NPN transistor will supply bias current to the PNP transistor. Once the PNP transistor is biased on, it will supply even more positive bias to the NPN transistor. This leads to a situation where both transistors are latched fully on and heavy current will flow through the device. Because of the base to emitter voltage drop of the two transistors, there will be a voltage drop of about one volt across the SCR when it is switched on. The SCR will remain in the latched on state until the anode voltage is reduced to about 1V and the current through the SCR falls below a threshold value, known as the holding current (Ih). The SCR is closely related to the Shockley diode and the PUJT (programmable unijunction transistor). The triac is another multilayer switching device which is quite similar to the SCR. The triac is a bi-directional device which is often used in AC switching and power control circuits. As the triac can conduct in either direction, the two main terminals are called MT1 and MT2 (main terminal 1 and 2) instead of cathode and anode. **Figure 4.30** shows the symbols for the SCR and triac.

The most common type of over-current protection circuit uses a resistor as the current sensing element. Once the voltage drop across the current sensing resistor reaches a predefined threshold, the output current of the supply is limited by reducing the output voltage. This type of current limiting prevents the external circuit from drawing excessive current. When this method of current limiting is used in a conventional linear power supply, it can result in very high power dissipation in the pass transistor. In the worst case scenario where there is a short circuit at the supply output, the output voltage will be close to zero. Under these conditions, the pass transistor will have to dissipate almost all of the power provided by the transformer for an indefinite period of time.

Many power supplies use a more sophisticated method of current limiting which is known as foldback current limiting or foldback protection. A power supply with foldback current limiting will deliver its full output voltage and current levels until the overload threshold is reached. Under overload conditions, the output current is reduced to a level which is much lower than the normal output current rating of the supply. This approach leads to a dramatic reduction in pass transistor dissipation.

My preferred method of current limiting is to shut down the power supply when the maximum current limit is exceeded. This type of protection is known as an over-current trip or electronic fuse. The circuit behaves like a conventional fuse or circuit breaker. Once the maximum current threshold is exceeded, the output voltage is reduced to zero. Under overload conditions, the output current is zero and no power is dissipated in the pass transistor. This means that the circuit can remain in the tripped state indefinitely. Most current-trip circuits will need to be manually reset once the cause of the overload has been removed.

THE CIRCUIT. We will build a simple trip circuit which uses a SCR as a latching switch. Since low power SCRs and programmable unijunction transistors like the BRY39 **[41]** are not so easy to find, our over-current protection circuit will use a pair of transistors as a home made SCR. The circuit is very similar to the example in Figure 4.29b. I used a BC548 for the NPN transistor and a BC558 as the PNP transistor. The choice of transistor is not very critical, many other general-purpose transistors would work just as well. **Figure 4.31** shows the schematic of the over-current trip circuit. The operation of this circuit is quite simple and easy to understand. The circuit is designed to work with the 13.8V power supply. The input marked D is connected to the source terminal of the MOSFET pass transistor. The voltage drop across the 0.1Ω resistor in the source circuit of the MOSFET is directly proportional to the output current of the power supply. With the component values shown in Figure 4.31, the protection circuit will be activated when the voltage at point D reaches 1.2 to 1.4 volts. When the protection circuit is triggered, the two transistors will conduct heavily. This will pull the gate voltage of the MOSFET pass transistor down to around 1 volt. Since this is less than the turn-on threshold of a power MOSFET, the pass transistor will be biased off and the output voltage of the power supply will quickly fall to zero. Once the current protection circuit is activated, it can only be reset by pressing the reset button or switching off the power supply until the main smoothing capacitor has discharged.

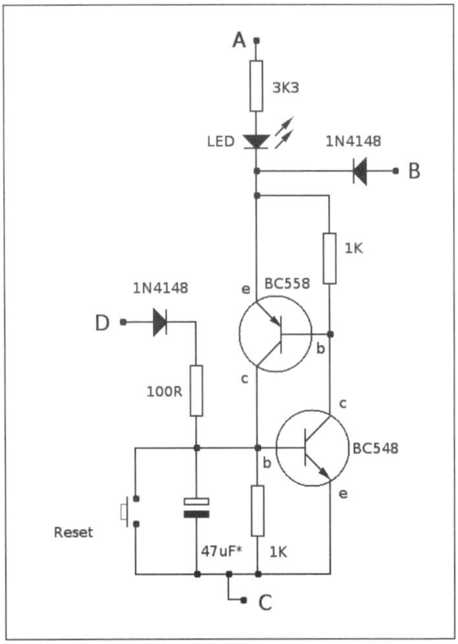

FIGURE 4.31: Overcurrent trip circuit.

Conventional fuses are relatively slow devices. Even fast-blow types can take seconds to blow if

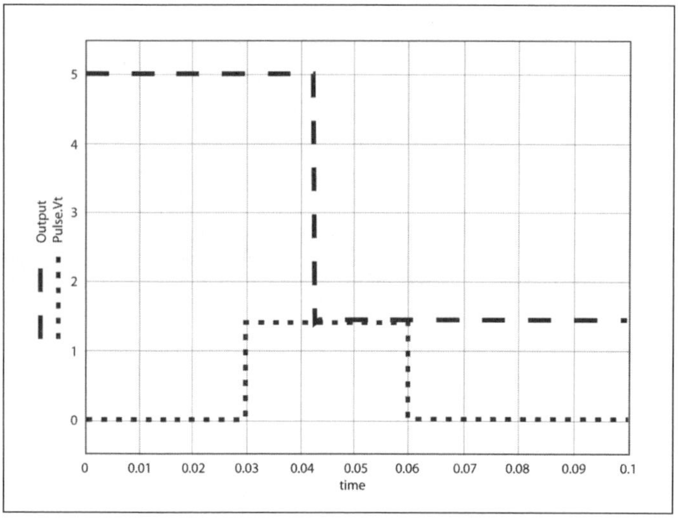

FIGURE 4.32: Quite Universal Circuit Simulator simulation of the circuit in Figure 4.31.

the current overload is less than several times the maximum current rating of the fuse. The electronic over-current trip circuit is much faster than even the fastest fuse. This can lead to problems when the power supply is feeding a capacitive load. The initial surge of current while the load capacitance is charging will tend to trip the protection circuit. Most over-current protection systems have a slight delay to reduce the risk of spurious triggering of the protection circuit. The 47µF capacitor in Figure 4.31 gives a delay of a few milliseconds before the over-current trip operates. **Figure 4.32** shows a QUCS (Quite Universal Circuit Simulator) simulation of the circuit. The dotted trace shows a 1.4V pulse which is applied to the current sense input (D). This is to simulate a 14A current surge for a period of 30ms. The dashed trace shows the voltage at point B in the regulator unit.

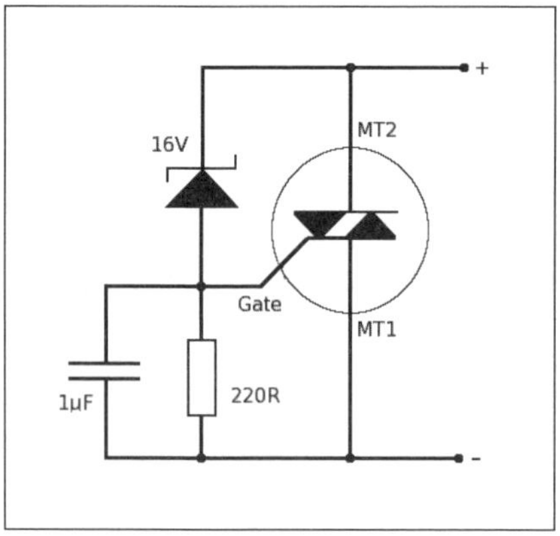

FIGURE 4.33: Circuit of the overvoltage trip.

OVER-VOLTAGE PROTECTION. The over-voltage protection circuit uses a simple and rather brutal method of protecting equipment that is connected to the power supply. This type of protection device is known as a crowbar circuit. If the output voltage of the power supply exceeds a predefined level, the crowbar protection will be activated and a high current SCR or triac will short-circuit the output of the supply. This will pull the output voltage down to around one volt and cause a huge current surge which operates the over-current trip or blows the fuse.

Most crowbar circuits use a Zener diode as the voltage sensing device and a SCR as the high-current crowbar. I used a TIC246D **[42]** triac instead of an SCR because high-power triacs are cheap and readily available. The TIC246D is a 16A/400V device. Although the RMS current rating of this triac is only 16A, it can withstand surges of 125A for long enough to operate the over-current trip.

CONSTRUCTION. **Figure 4.33** shows the schematic of the crowbar circuit. The circuit is so simple that I didn't bother to design a PCB layout for it. The triac is mounted on some copper PCB laminate. A SCR with similar maximum current ratings could be used in its place. The heatsink tab should be insulated from ground by a mica or silicone rubber insulator. The Zener diode is a standard 1.3W type with a Zener voltage of around 16V. **Photo 4.7** shows the layout of the crowbar circuit. The crowbar unit should be connected to the output terminals of the power supply using two short lengths of fairly thick wire (1mm or more).

The prototype regulator board was built on a strip of Veroboard. Since last month, I have designed a PCB layout for the regulator and current trip circuit. The board layout is shown in **Figure 4.34**. The PCB artwork can be downloaded from: http://tinyurl.com/2rxqpw.

PHOTO 4.7: The prototype crowbar circuit.

The power supply has been designed so that it can be easily scaled up for use at higher current levels. By using a bigger transformer, rectifier, smoothing capacitor and several MOSFETs in parallel, the maximum output current can be increased to several tens of amps or even hundreds of amps. **Figure 4.35** shows how to connect multiple pass transistors in a manner that won't interfere with the operation of the over-current trip circuit. Using a 0.1Ω

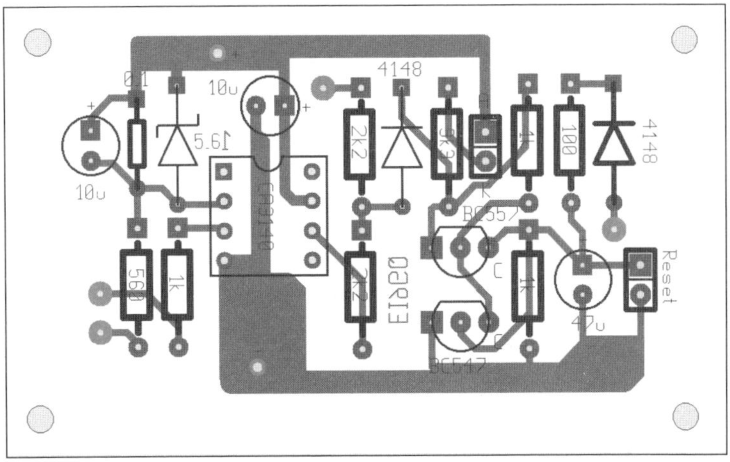

FIGURE 4.34: PCB overlay for the regulator and current trip circuit.

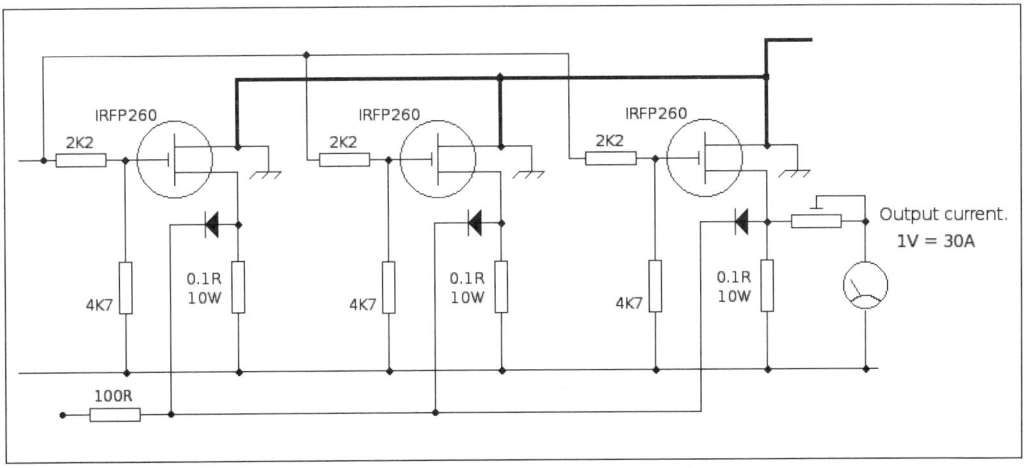

FIGURE 4.35: Connecting multiple pass transistors in series.

source resistor with each MOSFET will give a trip current of 12-14A multiplied by the number of pass transistors.

TESTING. The overcurrent protection can be tested by deliberately overloading the output of the power supply. I used another IRFP260 MOSFET which was mounted on a large heatsink as a dummy load for testing the power supply. Gate bias for the MOSFET in the dummy load was provided by a 10k pot which was connected between the drain and source terminals of the MOSFET. Car headlight bulbs also make a good dummy load for PSU testing. Note that the peak inrush current of a cold tungsten filament is much higher than the normal operating current of the bulb. Tests with the MOSFET dummy load proved that the protection circuit would trip at somewhere between 12 and 14 amps. Once I was satisfied that the over-current and over-voltage protection was reliable, I used my home made 100W SSB transmitter for further tests. At 50W PEP output, the transmitter draws peak current of less than 10A from the PSU. Increasing the transmitter output to about 65-70W will trip the protection circuit and shut down the supply.

I did several short-circuit tests by placing a metal rod across the output terminals of the supply. The power supply shut down immediately and didn't suffer any damage during the tests. I wouldn't recommend this method of testing because it might damage the smoothing capacitor or bridge rectifier.

There are two ways of testing the over-voltage crowbar. The crowbar can be tested separately using a variable-voltage/variable-current power supply before it is installed. Alternatively, the crowbar can be tested in the homebrew power supply by adjusting the output voltage control (4.7k multi-turn pot) until the output voltage reaches about 16.7V (Zener voltage + 0.7V). At this point, the crowbar will operate and the over-current circuit will shut down the supply. The current overload LED should indicate that the protection circuit has tripped. Once the circuit has been tested, the voltage should be adjusted back to 13.8VDC. **Photo 4.9** shows the completed PSU.

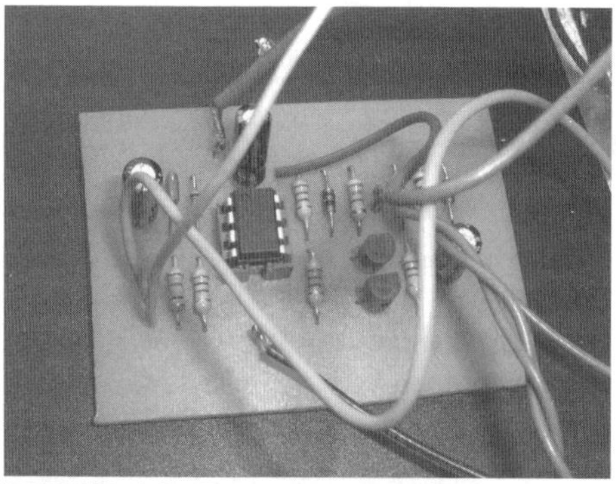

PHOTO 4.8: Prototype regulator and current trip circuit.

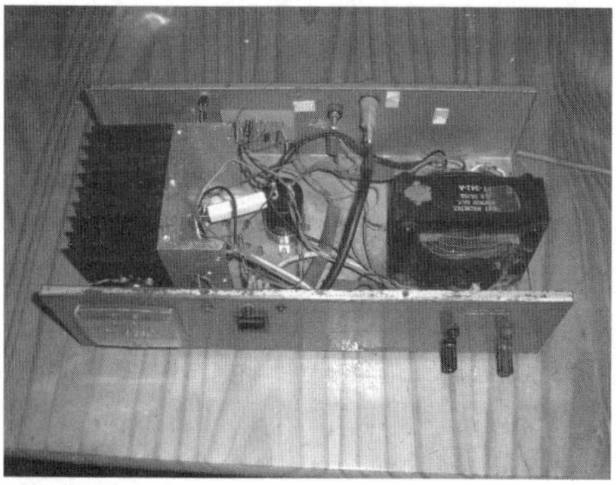

PHOTO 4.9: The completed prototype power supply unit.

The power supply has proved to be reliable over several months of testing. It seems to be quite immune to high levels of RF at HF and VHF frequencies. The output voltage regulation is unaffected by the close proximity my 250W HF and 300W VHF linear amplifiers. I chose the CA3140 op-amp for the voltage regulator stage of this project because of its ability to remain linear, even when the output voltage is close to the supply rails, and its very high input resistance. I have tested a few other types of op-amp in the regulator stage and found that the LF351 and even the humble xx741 will perform well in this circuit.

A VHF to HF Receive Converter

The minimum requirements for a VHF receive converter are a local oscillator and frequency mixer. Most 2m converter designs use an IF of 28-30MHz. The choice of IF is a trade off between local oscillator stability of the HF receiver and IF image rejection of the VHF converter. Since a 28MHz IF puts the image signal 56MHz away from the wanted signal, good image rejection is quite easy to achieve. 10m is also the only HF band that is almost as wide as the European 2m band. Although the current 10m amateur band only extends from 28.0MHz to 29.7MHz, most band-switched amateur receivers can be tuned from 28-30MHz which allows coverage of the entire 144MHz-146MHz 2m band.

Now that many amateur band receivers and transceivers offer general coverage reception from LF to 30MHz and beyond, it is possible to use some of the lower frequency HF bands as the IF. Some transverter designs like the state-of-the-art Javornik Transverter **[43]** use an IF of 14MHz. Frequency stability of the HF receiver is not such a big issue with modern receivers because they usually have crystal controlled PLL/DDS synthesisers as the local oscillator. The frequency stability of such receivers is almost as good at 30MHz as it is at 1MHz.

The most basic receive converter would consist of a band pass filter to remove the IF image and other out of band signals, a local oscillator and a frequency mixer. This simple converter would be quite adequate for listening to strong signals from local stations or FM repeaters. For reception of weak signals, a more elaborate circuit will be required. Most VHF converters will have at least one stage of RF amplification before the mixer. The mixer can be a passive diode mixer or an active mixer using one or more transistors or FETs. Passive mixers are often followed by a post-mixer amplifier. See the 80m receiver project for an example of a typical diode DBM and post-mixer amplifier circuit.

NOISE CONSIDERATIONS AT HF. On the lower frequency HF bands, the sensitivity of a receiver will tend to be limited by atmospheric and man-made noise. There is little point in using a very sensitive receiver when background noise levels are very high. I live in a typical suburban location where the noise level on 80m is always above S-3, even during daylight hours or in the early hours of the morning. The situation is much worse in the late afternoon and evening when the band is usually very busy. Quite often the noise floor is as high as S-9 which is about 100μV p-p at the input of the receiver. My main receiver is equipped with an RF preamp, but I never need to use it on any frequency below 14MHz. Much of the noise is from local sources like switching power supplies, TV timebase harmonics, spark generators for central heating boilers, car and motorcycle ignition systems and motors in washing machines and other domestic appliances. Due to the nature of HF propagation, noise generated by lightning can often be heard by stations that are hundreds of miles away from the thunderstorm. In addition to the local noise, we must also cope with QRM from distant stations (who never seem to be able to hear my weak signals) and splatter from strong stations on adjacent frequencies. It's a wonder that any of us ever manage to hear any DX stations on the lower bands!

NOISE CONSIDERATIONS AT VHF. The situation is quite different at VHF and above. Apart from the occasional shower of statically charged rain or snow, noise from atmospheric sources is rarely heard on VHF. Man made noise is also much less of a problem at VHF. Highly directional 'beam' aerials will give substantial gain in a favoured direction while suppressing noise and interference from all other directions. Even in this relatively noisy city location, the background noise on 2m is low enough that I can hear the increase in noise when I point my 9-element 2m Yagi directly at the sun. Against this very quiet background, it should be possible to resolve very weak sub-microvolt signals with a reasonably good signal to noise ratio (SNR).

NOISE FIGURE. All amplifiers generate noise. A very noisy amplifier will make a lot of noise, a very quiet amplifier will make less noise, but it will always make some. This means that there is a practical limit to the sensitivity of all receivers. No matter how much gain the receiver has, its sensitivity will be limited by the noise which is generated in the early RF amplifier stages. A perfect noiseless amplifier would amplify the input signal without adding any extra noise. The output signal to noise ratio would be exactly the same as the input signal to noise ratio. Unfortunately, such amplifiers do not exist in the real world. All amplifiers generate some noise and distortion. The output signal of an RF amplifier will be an amplified replica of the input signal with some unwanted noise and distortion added by the amplifier. The noise figure (NF) of an amplifier is a useful figure of merit for low noise amplifiers. The noise figure is the ratio of the input SNR to the output SNR expressed in decibels. Note that 'noise figure' is not the same as 'noise factor'. Noise factor is the ratio of the input SNR and output SNR as a simple power ratio. Noise figure is the same ratio measured in decibels.

Most modern low noise amplifiers use FETs (field effect transistors) as the active device. The standard silicon junction FET has largely been replaced by special low noise devices like the dual-gate MOSFET, GaAsFET (Gallium Arsenide FET) and HEMT (high electron mobility transistor). Since the late 1980s, the demand for low cost, high performance transistors for use in satellite TV receivers has led to the development of very low noise devices which are readily available and cost no more than a few pounds per transistor.

A good VHF pre-amplifier will have a noise figure of around 1dB. Very low noise amplifiers can achieve noise figures of 0.5dB or even less at room temperature. Even lower noise figures can be achieved by cryogenically cooled amplifiers as used in radio telescopes. State of the art LNAs use HEMT devices which are cooled to within a few degrees of absolute zero. Some of the more esoteric designs use matching networks and filters which are made from high temperature superconductors. 'High temperature' in this context means higher than absolute zero but still very, very cold. Radio telescopes can take advantage of this level of amplifier performance because they use very high gain aerials which are directed away from sources of thermal and man-made noise. Very few radio amateurs would have any need for such extreme low-noise reception techniques. Fewer still would be willing to pay the very high cost of a refrigeration system to cool the front-end amplifiers.

PRACTICAL RF AMPLIFIERS. FETs and bipolar transistors can both be used as RF amplifiers. The latest generation of UHF and microwave bipolar transistors offer a good compromise between dynamic range and low noise. GaAsFET and HEMT devices as used in satellite LNBs are an excellent choice for use as VHF/UHF preamps. High power GaAsFETs which are designed for use in UHF and microwave transmitters can make extremely good receive RF amplifiers. Dual-gate MOSFETs like the venerable BF981 and newer surface-mount versions like the BF996S make very good RF amplifiers for frequencies up to about 200MHz. With careful input matching, a BF981 can deliver about 25dB of gain and a noise figure of around 1dB at 144MHz. This level of performance is suitable for all but the most demanding applications. I have heard EME (moonbounce) signals on 2m using a 9-element Yagi and a simple receive converter with a dual-gate MOSFET as the RF amplifier.

Dual-gate MOSFETs and GaAsFETs will typically have a gain of somewhere between 20dB and 30dB at 144MHz. This means that losses and noise at the amplifier input will have a significant effect on system noise figure. The amplifier output circuit is much less critical. The noise performance of the entire receiving system will tend to be dominated by the performance of the first stage. Amateur constructors go to great lengths to reduce losses at the input of low noise amplifiers. As the noise figure of modern transistors can be as low as a fraction of 1dB, even tiny losses in the input matching network will have a significant effect on the overall system noise figure. VHF designs often use silver plated copper wire for the inductors in the input matching network. Capacitors used in the input network should be high-Q, low-loss types. UHF and microwave preamps often use resonant lines and cavities made from silver plated metal. Some 23cm preamps are as much a work of art as they are a work of engineering **[44]**.

Any of the matching networks described earlier in this chapter can be used to match the feedline impedance (usually 50Ω) to the input of the RF transistor. L, PI, T, CLC and other controlled-Q networks are all quite suitable for this job. **Figure 4.36** shows a few commonly used matching networks. **Figure 4.36A** is a T-network, B is a CLC network as used in our 10m PA circuit, C is a simple parallel tuned circuit with capacitive coupling to the 50Ω feedline. The number of components in the input circuit should be kept to a minimum. Unnecessary components can lead to increased losses and may

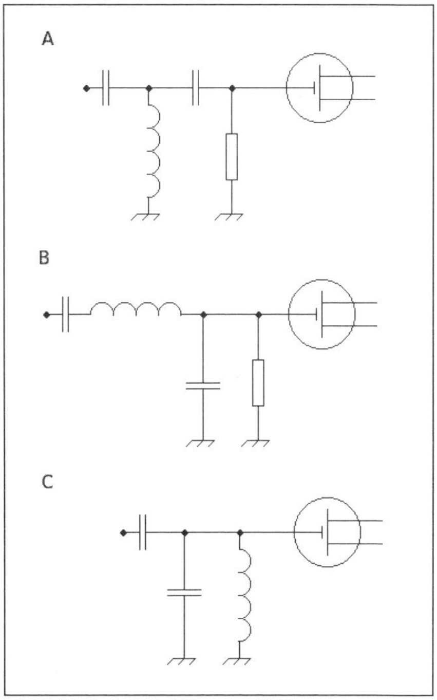

FIGURE 4.36: Three different methods of input matching (see text).

be a source of noise. A typical preamp will have a low-loss connector for the aerial feedline. Good quality SO239 connectors are just about ok for 2m, N-type or SMA connectors are preferred for 70cm and higher bands. The matching circuit at **Figure 4.36C** has a slight advantage over the other two configurations. As the inductor provides a DC path to ground for the gate bias supply, there is no need for an additional choke or resistor as required in circuits A and B. The inductor should be wound from fairly thick copper wire. A coil wound from silver plated wire will have higher unloaded Q. The capacitors should be high quality, preferably air-spaced or brass/ceramic piston types. Where minimum losses are more important than small size, the inductor can be replaced by a tuned line and cavity. The line can be a flat stripline or a length of metal tubing. Our local 2m moonbounce guru uses 15mm copper water pipe as part of the input matching network of his 2m preamp!

THE CIRCUIT. All of the modules in the transverter will be built using dead-bug style construction. Note that this transverter is a work in progress rather than a retrospective look at a completed project. The modules are designed to be easily modified or replaced. My old transverter was built in 2002. Since that time, the original receive converter section has been replaced/upgraded twice. The new transverter will probably suffer the same fate.

The local oscillator is a simple common-collector Colpitts crystal oscillator. The local oscillator frequency is 144-28=116MHz. The oscillator runs at 38.6666MHz, which is one third of the final local-oscillator frequency. The crystal is a third overtone type. This 38.6666MHz signal is tripled to 116MHz in the collector circuit of the oscillator transistor. A dual-gate

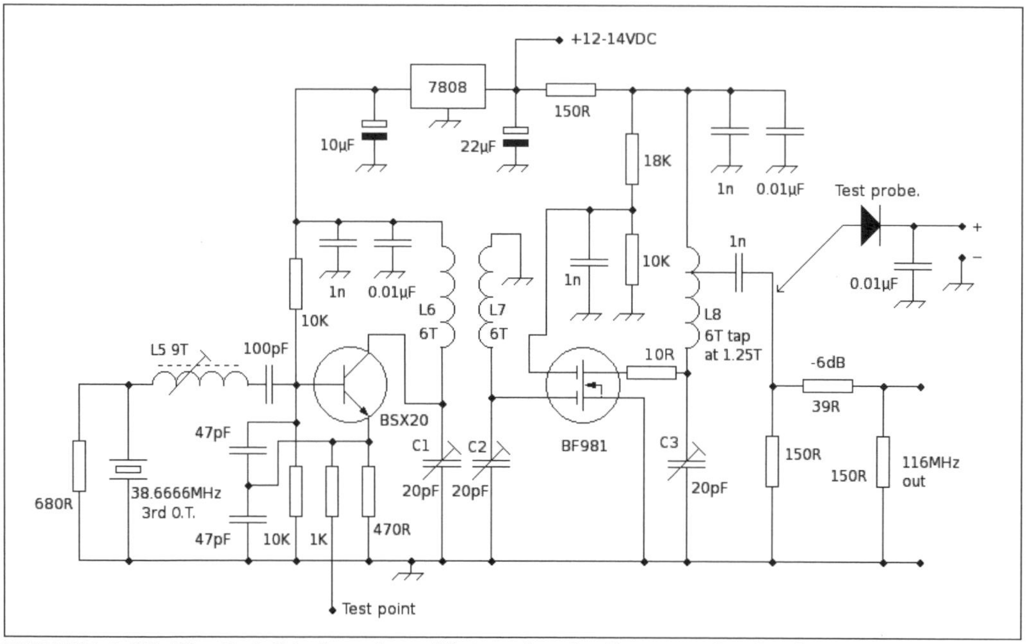

FIGURE 4.37: The Colpitts-based oscillator circuit. See text for further coil details.

MOSFET buffer amplifier gives excellent isolation of the oscillator and increases the local-oscillator signal to a level of several volts p-p. An 8 volt regulator IC provides a clean and stable power supply for the oscillator. The local-oscillator module can deliver more than 5V p-p into a 50Ω load. We don't need this much local-oscillator power for the FET mixer in the receiver converter section, but it will come in handy when we get around to building the transmit mixer section. A 6dB attenuator is used to reduce the LO module output to a suitable level. Component values for the attenuator are shown in the schematic. L6, L7, L8 and the associated trimmer capacitors act as a band-pass filter that suppresses the oscillator fundamental frequency and any unwanted harmonics.

The receive converter module uses a dual-gate MOSFET as the RF amplifier and a JFET as the 144-28MHz frequency mixer. The 50Ω input is matched to gate-1 of the MOSFET by a simple matching network of the type shown in Figure 4.36c. The network is designed to have a relatively low loaded Q to keep losses to minimum. The RF amplifier output is fed to the mixer via a band-pass filter. The mixer is a simple, single-ended JFET mixer. The high output impedance of the mixer is matched to the 50 ohm input of the HF receiver by a PI network. Gate-2 bias for the MOSFET is provided by a simple resistive voltage divider.

CONSTRUCTION. All of the air-cored inductors are wound from 1mm copper wire. I used the 1mm centre core from a length of 75Ω TV co-ax. A 5mm drill bit was used as a former for winding the coils, giving a centre-to-centre coil diameter of 6mm. L6, L7 and L8 in the oscillator module are all 6 turns with a fairly close turns spacing. Each turn is spaced by slightly less that the wire diameter. L8 has a tap at 1.25 turns from the DC end for the 50Ω

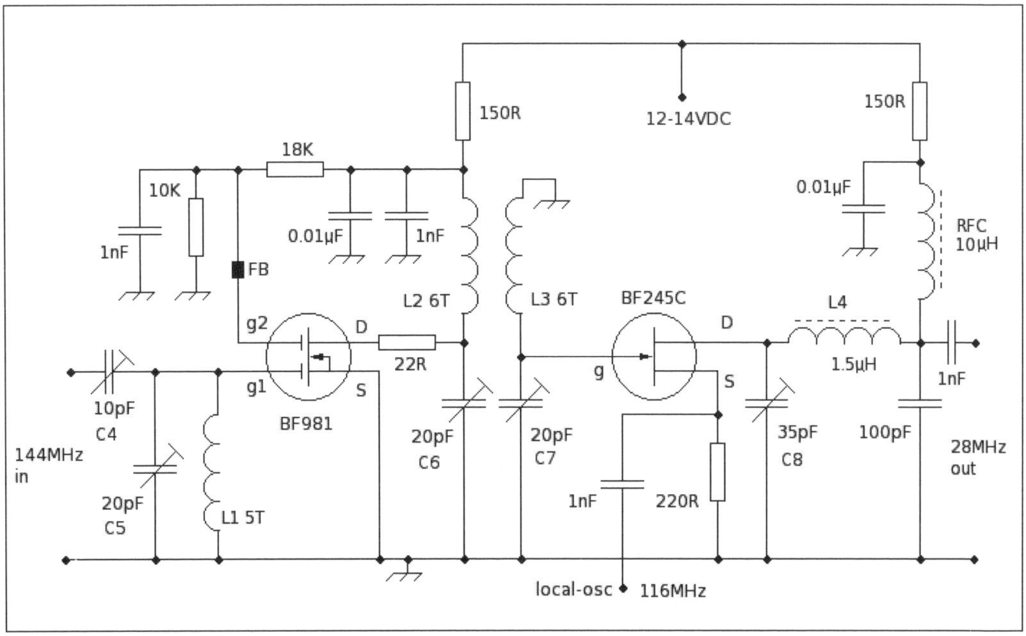

FIGURE 4.38: The preamplifier and receive down converter circuit.

output. L5 in the oscillator module is 9 turns on a small slug-tuned former. I used a 5mm former, with a powdered-iron tuning slug. If you use a different former, you may need to adjust the number of turns. L1 in the converter module is 5 turns with a turn spacing of slightly more than the wire diameter. L2 and L3 are 6 turns each with a wire spacing of slightly more than the wire diameter. To minimise stray coupling between the input and output circuits, L1 is kept as far as possible from the coils at the output of the amplifier. L1 is positioned at 90 degrees relative to the output coils. See **Photo 4.10** for details. L4 in the mixer output network is 20 turns on a T50-6 yellow powdered-iron toroid. Dead-bug construction allows for easy modification of the receive converter. It would be quite easy to replace the BF981 with a surface-mount device like the BF996S or a dual-gate GaAsFET like the 3SK124 or 3SK97. GaAsFET devices need a lower supply voltage than the dual-gate MOSFET, so you will need to use a Zener diode or voltage regulator IC. The local-oscillator module produces more than enough output to drive a high-level diode mixer. If you have such a mixer in the junk-box, why not try it in this circuit? Even if you don't have one, you could always make your own diode mixer. Don't be afraid to experiment, if you are not happy with the changes you make, it won't take more than a few minutes to revert to the original circuit.

TESTING. The local oscillator should be built and tested separately before it is connected to the converter module. Connect the local-oscillator unit to a 12V power supply. If you have a frequency counter, you should connect it to the test point shown in the schematic and adjust L5 until the oscillator frequency is 38.666667MHz. If you don't have a suitable frequency counter, you can use an off-air signal as a frequency reference. The carrier frequency of the GB3VHF beacon on 144.430MHz is a very accurate frequency reference.

Next, connect a DC volt-meter to the test point. Slowly tune trimmer capacitor C1 until you see a dip in the voltage at the test point. The dip indicates that the collector circuit is tuned to a harmonic of the 38.666MHz oscillator. I find that this dip occurs when the plates of the 20pF capacitors are about 30-40% meshed. Once you have found the correct position for C1, you should adjust C2 and C3 to the same position. This may not be the correct position for C2 and C3, but it should be a reasonably good starting point for further adjustments. I used a simple diode probe as shown on the schematic for measuring the output of the oscillator module. Connect the diode probe to the point indicated on the schematic. A DC volt-

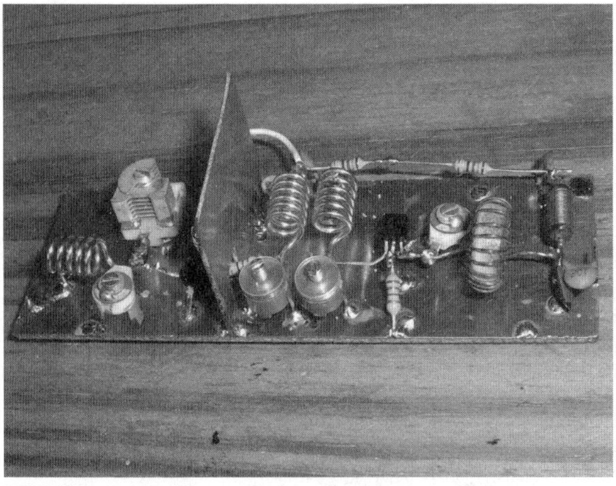

PHOTO 4.10: The receive down converter complete with input-output screen (see text).

PHOTO 4.11: Practical realisation of the oscillator.

meter connected to the diode probe will indicate the peak output voltage of the oscillator unit minus the voltage drop of the diode. A germanium point-contact diode or a Schottky signal diode is ideal for this purpose. Adjust C2 and C3 for maximum output. There is some interaction between the tuning of L5 and C1. Once C1, C2 and C3 are properly tuned, you may need to make a slight adjustment to L5 to get an output frequency of exactly 116.0000MHz.

As I have previously built several RX converters using dual-gate MOSFETS, I was quite sure that this one would work perfectly as soon as I turned the power on. I was rather disappointed to discover that the RF amplifier stage was oscillating at some unknown frequency. A loud rushing noise could be heard from the HF receiver. When I waved my hands near the RF amplifier, I could hear all kinds of nasty whistling noises from the loudspeaker! I tried a few of the standard remedies for unstable MOSFETs to see if I could cure the problem. I put a small ferrite bead on the G2 lead of the BF981 and increased the value of the resistor in the drain circuit from its original value of 10 to 22Ω as shown in the schematic. Still no joy! Next, I made a shield from double-sided PCB laminate and placed it between the amplifier input and output circuits. See **Photo 4.10** for details. This cured the problem completely. The amplifier is now unconditionally stable.

The receive converter module can be aligned with the help of a signal generator or a strong off-air signal. Adjust C4 so that the plates are about 30% meshed. Adjust C5, C6 and C7 so the plates are about 40-50% meshed and C8 until the plates are about 70% meshed. Use a signal generator or a strong 2m signal to tune C5, C6, C7 and C8 for best reception. Once C6, C7 and C8 are tuned properly, they should not require any further adjustment. Try to find a very weak signal like a propagation beacon or a distant repeater and tweak C4/C5 for best signal to noise ratio. If you are using a HF transceiver as the 10m receiver, make sure you don't operate the transmitter. A few watts of RF would probably destroy the JFET mixer.

When I connected my homemade 9-element Yagi to the converter input, I was able to hear the signals from GB3VHF quite clearly. Despite poor tropospheric conditions and low barometric pressure, I have been able to hear GB3VHF every day since I built the converter. My location is in grid square IO51TV which is more than 600km away from the beacon transmitter. A recording of my first test can be found at http://tinyurl.com/2pxaq9

Transmit Converter for the VHF Transverter Project

We will use the local oscillator module from the VHF receive converter to provide the 116MHz local oscillator signal for the transmit converter. As the oscillator module is capable of providing sufficient output to drive both the receive and transmit mixers, there is no need for any complicated RX/TX switching of the local oscillator signal. A Wilkinson **[45]** power splitter will be used to divide the local oscillator power equally between the two mixers.

CO-AXIAL TRANSMISSION LINE TRANSFORMERS. Most aerial experimenters will be familiar with the practice of using co-axial or open wire transmission line as an impedance transformer. Various combinations of series or parallel connected transmission lines can be used as an impedance matching device. Open circuit or short circuit transmission lines can also be used as part of filter networks or as the resonant circuit in an oscillator. At UHF and microwave frequencies, the resonant lengths of transmission line are small enough to be etched on double sided or multi-layer PCBs.

One of the most common transmission-line transformer arrangements uses a series section made from a quarter-wave length of transmission line. Simple unbalanced to unbalanced transformers can be made from co-axial cable or printed PCB microstrip line. For a typical application where we want to make an impedance transformation from Z1 to Z2, the characteristic impedance of the quarter-wave matching section should be $\sqrt{(Z1*Z2)}$. **Figure 4.39A** shows a quarter-wave matching section made from co-axial cable. Note that the line is an electrical quarter-wavelength which is not the same as a free-space quarter-wavelength. As we will be using a pair of quarter wave lines to split the local oscillator signal, each line will need to transform 100 to 50Ω . The two 100Ω ports are connected in parallel to make 50 ohms. $\sqrt{(50x100)} = 70.710678\Omega$. A transformer made from a transmission line with this exact value of characteristic impedance would have ideal characteristics at its design centre frequency. A QUCS simulation of such a transformer shows a return loss of well over 100dB and through loss of 0dB. Unfortunately, we usually have to make do with a less ideal situation. 70.7Ω is not a standard value for co-axial cable, so we will have to use the nearest available standard value. 75Ω co-ax is widely available and quite suitable for use in a 50Ω power splitter. Unlike the ideal transmission-line in our QUCS simulation, the real 75Ω co-ax will have a small power loss. A properly terminated quarter wave transformer made from 75Ω co-ax will have a return loss

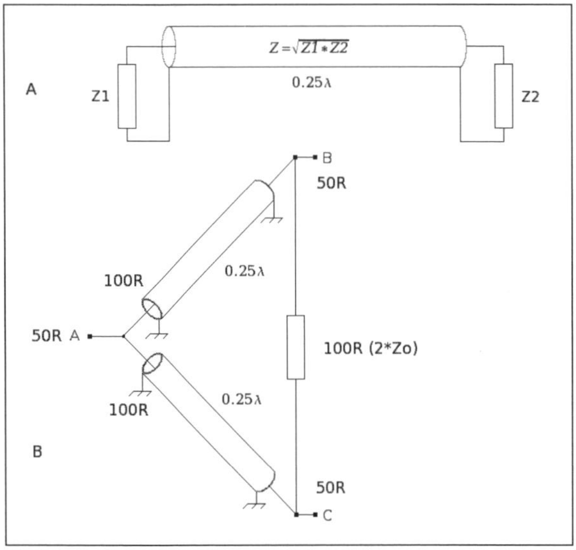

FIGURE 4.39: Quarter wave matching line and splitter circuit.

of 20-25dB at both ends and through loss of a small fraction of a decibel.

THE WILKINSON POWER SPLITTER/DIVIDER. The power divider circuit shown in **Figure 4.39B** uses a pair of 100 to 50Ω transformers to divide a signal applied to port A equally between ports B and C. If you remove the resistor between port B and C, you are left with the classic two-way splitter/divider, which is often used to feed VHF/UHF aerial arrays. This simple splitter can be used to feed two aerials, and more splitters can be added to port B and C to feed arrays of 4, 8, 16 or even more aerials. Adding a resistance of 2*Zo (100Ω) between ports C and D has an interesting effect on the performance of the circuit. This creates two paths between B and C, one through the resistor and the other through the two quarter-wave lines. Because the path via the transmission lines causes a phase shift of 180° degrees, a signal applied to port B will be cancelled out at port C and a signal applied to port C will be cancelled out at port B. This port-to-port isolation is one of the most useful features of the Wilkinson power divider. When a signal is applied to port A, it is divided equally between ports B and C. Because the signals at B and C are in phase, no power will be dissipated in the resistor, provided that all three ports are properly terminated. **Figure 4.40** shows the physical layout of a UHF or microwave power splitter. The two quarter-wave lines are printed microstrip lines on double-sided PCB. The resistor is a low inductance surface mount type.

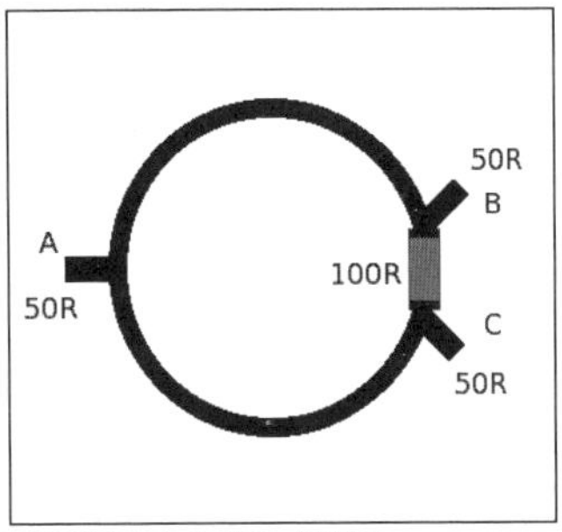

FIGURE 4.40: UHF or microwave power splitter using printed quarter wave lines.

Circuits based on co-ax or microstrip lines are widely used at UHF and higher frequencies. Such circuits are usually impractical at HF and the lower VHF bands because of the size of the transmission lines. Conventional inductors and capacitors (lumped elements) can be used to make a circuit which is equivalent to a quarter-wave transmission line. **Figure 4.41A** shows an unbalanced coaxial transmission line. **Figure 4.41B** is a balanced transmission line. This type of line is known by various names like: flat twin, balanced feeder, open-wire line and ribbon feeder. The characteristic impedance of the line is determined by the inductance of the wires in the line and the capacitance between the two (or more) wires. When losses in the line are so low as to be negligible, the characteristic impedance of the line is $Zo=\sqrt{(L/C)}$ where Zo is the characteristic impedance in ohms, L is the line inductance

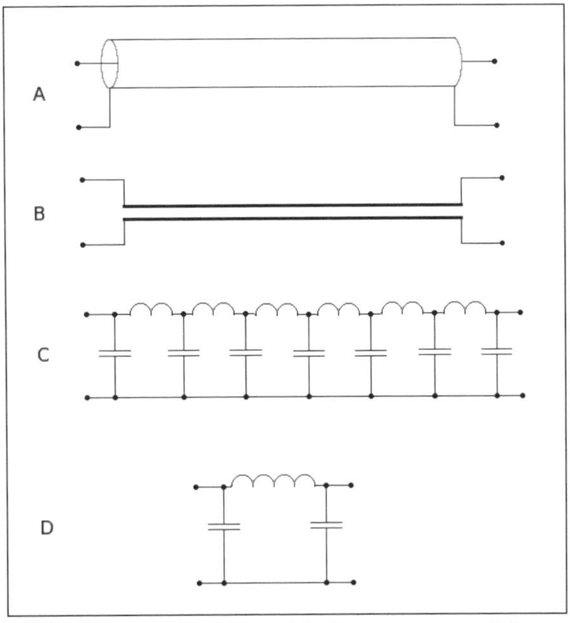

FIGURE 4.41: Relationship between coaxial line and its lumped-constant equivalent.

per unit length and C is the capacitance per unit length. The distributed inductance and capacitance of an unbalanced line is shown in **Figure 4.41C**. These distributed reactances can be replaced by lumped reactive elements as shown in **Figure 4.41D**. You will recognise this circuit as a PI network as used in several of our previous Homebrew projects.

Figure 4.42 shows a Wilkinson divider made from lumped circuit elements. The capacitors are ordinary ceramic types and the inductors are air-cored copper coils. The component values were calculated using a standard formula from the VHF/UHF Manual **[46]**. My first attempt at running a QUCS simulation of the circuit shows that our power splitter has ideal characteristics. Return loss and port B/C isolation are almost infinite and through loss is 3dB from port A to either output port. This is a result of using the exact calculated values for all components, a perfect 50.000Ω termination on all ports and using a value of PI which is accurate to 10 decimal places. For the second attempt, I substituted the nearest standard values for C1 and C2, rounded the value of L to the nearest nH and added a small loss resistance to each component. Tests on the assembled circuit show that this model is reasonably accurate. A QUCS power vs frequency plot is shown in **Figure 4.43**. The dashed trace shows power loss from port A to port B (or C). The solid trace shows port B/C and C/B isolation and the dotted trace shows return loss at port A.

The Wilkinson divider/combiner is a narrowband device. Good isolation and low return-loss can only be achieved over a limited bandwidth. The circuits in Figure 4.39 and Figure 4.42 work well over a bandwidth of about 10% of the centre fre-

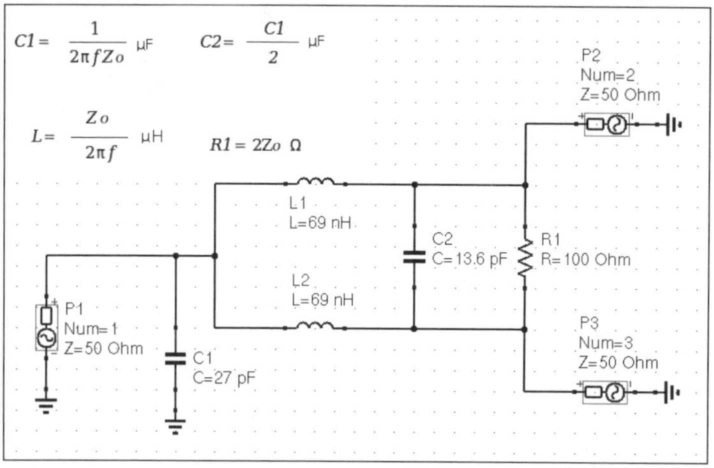

FIGURE 4.42: Wilkinson power divider made from lumped circuit elements.

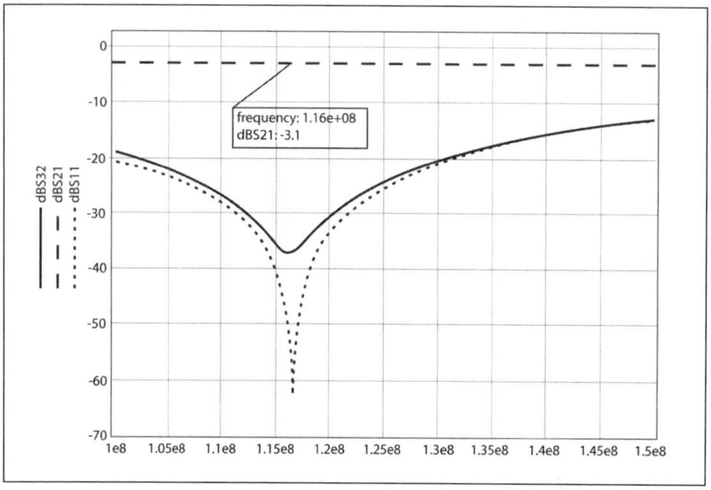

FIGURE 4.43: QUCS power vs frequency plot. Red trace: power loss from port A to B (or C). Green trace port B/C and C/B isolation. Blue trace: return loss at port A.

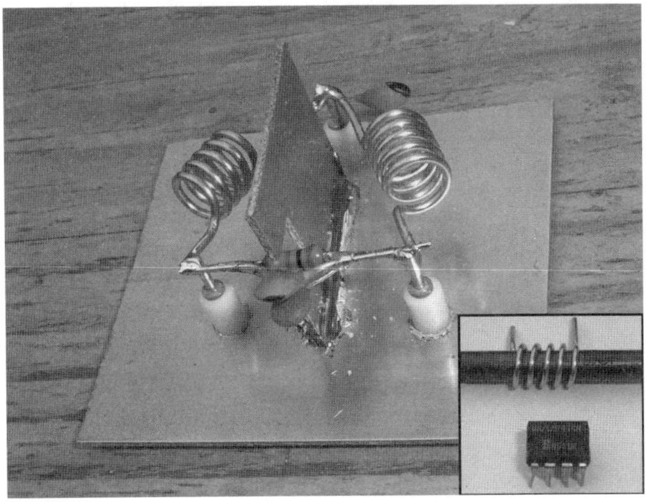

PHOTO 4.12: The completed Wilkinson power divider. Inset winding the coil - IC for scale.

quency. This is sufficient to cover any single amateur band. Dividers made from co-axial lines have an advantage over lumped element types because it is very easy to arrange for good screening between the two arms of the divider. Dividers made from printed microstrip lines can also offer excellent port-to-port isolation if stray coupling between the two quarter-wave striplines is kept to a minimum. The two arms of the divider network must be well separated, but the connections from port B and C to the isolation resistor must be as short and direct as possible. The circular stripline arrangement in Figure 4.40 provides a good solution to these conflicting requirements. Dividers built from lumped components usually have a metal screen between the two inductors to minimise stray coupling.

SPLITTER CONSTRUCTION. The power divider was built using point to point wiring on some copper-clad PCB laminate. Since there is no provision for fine-tuning of component values, the construction of the circuit is quite critical. The two inductors are mounted on Teflon stand-off terminals. Ceramic, glass or good old-fashioned tagstrip terminals can be used instead. The wire is 1mm bare copper. The centre conductor from standard 75Ω TV co-ax is ideal for this job. A 5mm drill was used as a former for winding the coils. This results in a 6mm centre-to-centre coil diameter. Each inductor is 5 turns with a turn spacing of slightly more than the wire diameter. The coil length is 12.5mm. See **Photo 4.12** for details.

The shield between L1 and L2 is made from double-sided PCB. C1 is a 27pF ceramic disc and C2 is a pair of 6.8pF capacitors connected in parallel. The capacitors should have a tolerance of 5% or better. R1 is a 100Ω 0.25 watt carbon-film or other low inductance type.

TESTING. The 116MHz local oscillator module can be used as a signal-generator for testing the power splitter. Connect the output of the LO module to port A of the power splitter. Port B and C should be terminated with a 50Ω resistor. You can use a pair of 100 ohm resistors in parallel or the nearest standard value of 51Ω. Use an RF volt-meter or a simple diode probe to measure the voltage across the two 50Ω resistors. The voltage at port B should be within a few percent of the voltage at port C. To test port B/C isolation, move the 50Ω termination from port C to port A and feed the LO signal to port C. You should find that the voltage at port B is now reduced by more than 20dB.

It is possible to make a broadband power divider/combiner. One of the most common types of broadband power splitter/combiner is based on a resistive bridge and ferrite core, broadband transformer. This circuit is known as a hybrid combiner, hybrid splitter or hybrid coupler. We will be taking a closer look at broadband divider/combiner circuits in a future Homebrew.

28/144MHz TRANSMIT CONVERTER. This is not really intended as a first project for the novice constructor. The design, construction and proper alignment of a VHF transmitter requires a great deal of care. Unless you have some experience of radio construction or you have access to a spectrum analyser, you should seek help from a more experienced constructor or from your nearest amateur radio club. Many clubs have test equipment which can be used or borrowed by club members. I hope this warning won't discourage newcomers from building VHF equipment. It is quite possible to build a VHF transmitter that is at least as good as the best commercial rigs.

Proper gain distribution and good filtering is the key to a successful transverter design. The mixer should always operate well within its linear region. We should always resist the temptation to run the mixer at higher levels in an effort to reduce the post-mixer gain requirements. This is a false economy because gain is cheap and easy to provide. Because of the harmonic relationship between many of the amateur bands, harmonics of the signal from the HF exciter will often fall within or very close to the transverter's output frequency. These harmonics can not be filtered out by band-pass filtering. Again, the best way to avoid this problem is to keep the HF drive at a low level. The use of a balanced or double-balanced mixer will help to suppress the odd harmonic frequencies (3rd, 5th, 7th and so on).

For example: if the transverter is operating at 144.3MHz and the HF exciter is tuned to 28.3MHz, the fifth harmonic will be at 141.5MHz. Using a balanced mixer will improve the suppression of this spurious response by 30 to 60dB compared to a simple unbalanced mixer. When designing and building a transverter, you should give careful consideration to the choice of IF and local oscillator frequencies. The transverter described here has an IF frequency of 116MHz which falls within or close to the aeronautical VHF band. The IF image frequency is at 116 – 28 = 88MHz which is at the bottom of the FM broadcast band. Since these two frequencies are not close to the 2m band, they will be well suppressed by the BPF. However, if you are very close to an airport, it might be wise to use an external BPF or co-ax stub type of band-stop (notch) filter.

The transmit converter takes a low-level signal from a 10m band transmitter and converts it to VHF. The converter mixes the 28MHz signal with the 116MHz local-oscillator to produce a 144MHz output. The mixer acts as a linear translator, which means that it can be used for SSB/CW/FM/AM/FSK or any other mode that is generated by the 10m transmitter. The output from the local-oscillator is divided equally between the transmit and receive mixers. The 28MHz signal from the 10m transmitter or transceiver is fed to the IF port of the mixer. A 50Ω attenuator is used to reduce the 28MHz signal to a suitable level for driving the mixer. The transmit mixer is a very critical part of a VHF transverter. Most transverter designs use a high-level balanced or double-balanced diode mixer. The mixer is followed by a band pass filter which suppresses the IF image and other out of band spurious signals. It is very important that this filter is correctly aligned and that the mixer is not overdriven. Failure to observe these simple precautions is likely to result in unacceptably high levels of spurious signals at the transverter output. The schematic of the mixer and BPF is shown in **Figure 4.44**.

Some HF rigs have a separate low-level output for driving a transverter. The value of input attenuation will depend on the amount of 28MHz drive available from the HF exciter. The transverter output from my rig puts out about 200mW. To make sure that the mixer IF port will never see more than about 1mW of drive, I need to use at least 23dB of attenuation at the mixer input. You may need to change the attenuator resistor values to match the output from your rig. You should avoid using a high-power HF transmitter with the drive control set for very low output for several reasons. Many HF transmitters are designed to give best linearity at their normal output power. Such transmitters may produce a very distorted SSB signal when they are operated at a small fraction of their normal output power. The drive or RF output control of most SSB transmitters usually controls the output of the low-level RF stages. The PA, driver and probably some of the pre-driver stages will still operate at maximum gain, even when the drive control is close to its minimum setting. Broadband noise produced by the PA and driver stages will be at a relatively high level compared to the SSB output. This noise will be converted to VHF by the transverter and could cause interference to other band users. The other more obvious problem with using a high-power transmitter is the increased danger of overdriving the transverter. Severe overdriving would lead to distortion of the VHF signal, increased spurious output and possible damage to the input attenuator or mixer.

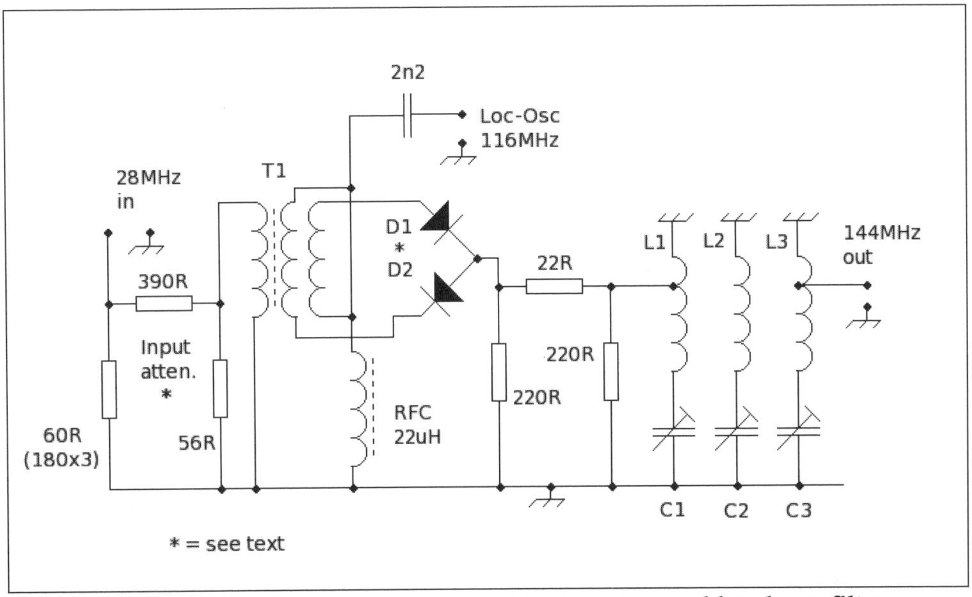

FIGURE 4.44: Circuit diagram of the transmit converter and bandpass filter.

CONVERTER CONSTRUCTION. Construction of the mixer is fairly straightforward. The circuit is built on PCB laminate using point-to-point wiring and Teflon stand-off insulators. T1 is wound on a small ferrite toroid or binocular balun core. The windings are 4 turns, trifilar wound. I used 1N5711 (5082-2800) Schottky diodes for D1 and D2. A pair of well matched 1N4148 or 1N914 diodes will perform almost as well. The band pass filter consists of three parallel LC tuned circuits. C1, C2 and C3 are 20pF trimmer capacitors. L1, L2 and L3 are each 6 turns of 1mm copper wire. A 5mm drill was used as a former for winding the coils. Spacing between each turn is equal to the wire diameter. The gap between each coil is about 2.5mm. L1 and L3 are tapped at slightly more than one turn from the ground end for the 50Ω I/O connections. **Photo 4.13** shows the prototype BPF and **Figure 4.45** shows the frequency response of the filter.

PHOTO 4.13: The completed three element filter.

TESTING AND ALIGNMENT. When I tested the receive converter last month, I installed a 6dB attenuator at the output of the local oscillator module. Now that the available local-oscillator power is to be shared between two stages, the attenuation at the local oscillator output should be reduced from 6dB to about 4dB. The two 150Ω resistors in the output attenuator should be replaced with 220Ω resistors and the 39Ω resistor should be reduced to 22Ω.

Final testing and alignment will be carried out when the transmit amplifier stages have been built and tested. If you have a VHF receiver, you can use it to do the initial alignment of the TX converter module. Connect a 0.5m length of insulated wire to the 144MHz output of the transmit converter module (tap on L3). Connect the local oscillator module to a 13.8V DC supply. Connect the output of the LO module to port A of the power splitter. Connect port B of the splitter to the LO input of the transmit mixer. Port C of the splitter should be terminated by a 50Ω resistor (or the nearest standard value).

Connect the transverter output of your 10m transmitter to the 28MHz RF input of the

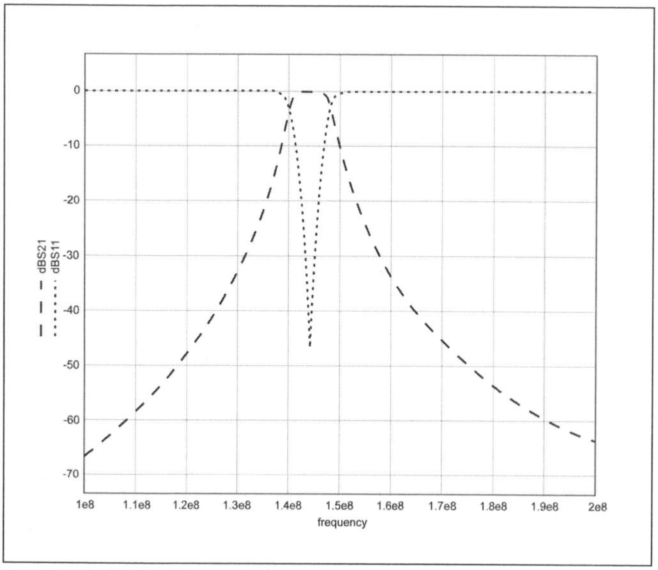

FIGURE 4.45: Plot of the return loss (dotted) and transmission (dashed) of the filter.

converter module. Tune the 10m transmitter to 29.000MHz and tune the VHF receiver to 145.000MHz. Adjust C1, C2 and C3 to approximately 50% of their maximum value (plates half meshed). When you operate the 10m transmitter, you should be able to hear the VHF signal at 145.000MHz. The power output from the BPF is approximately 100 microwatts, so don't expect a very strong signal. Adjust C1, C2 and C3 to peak the received signal. Tune the 10m transmitter to 28.000MHz and the VHF receiver to 144.000MHz and repeat the adjustment of C1, C2, C3. Finally, tune to 29.700MHz/145.700MHz and check the output signal. You should find that it is possible to get a flat response from 144MHz to 146MHz. The SSB output from the BPF should be clear and free from distortion. Tune the VHF receiver to check for distortion products (splatter) on adjacent frequencies. When the converter is driven at the correct level, you should find no sign of any spurious signals at any frequency beyond ±3kHz from the operating frequency.

PHOTO 4.14: A view of the completed transmit mixer with bandpass filter.

Power Levels and Gain Distribution in the Transmit Stages

POWER LEVELS AND GAIN DISTRIBUTION. The measured local oscillator injection level to the transmit mixer is +13dBm (20mW). We can expect this mixer to have an output 1dB compression point of about +7-10dBm and an output 3rd order intercept point (IP3) of +15-20dBm. As the 28-144MHz conversion loss of the transmit mixer is about 5-6dB, the input IP3 should be between +20dBm and +26dBm.

My choice of resistor values for the 28MHz input attenuator gives about 27dB attenuation. As the maximum saturated output of my 10m transmitter is 200mW, the highest possible drive level to the mixer will be 0.4mW (-4dBm). Under normal operating conditions the SSB drive level will be below 0.2mW (-7dBm). At this level, 3rd order IMD products at the mixer output will be 50-60dB below the wanted output and higher order IMD products should be so low as to be undetectable. Listening to the mixer output with a 2m receiver confirms that the SSB signal is of very good quality. The mixer is followed by a 4dB attenuator and a band-pass filter which loses another 1-2dB. The 144MHz power level is a mere 0.02mW or, if you prefer, 20μW. To bring this signal up to a useful level, we will need a lot of gain! I will be using this transverter to drive my valve linear amplifier. The linear amplifier needs about 5W of drive to give an output of 200W. This means that the transmit amplifier stages will need to have a power gain of at least 54dB. To achieve this much gain at VHF would normally require five stages of amplification. If we use a high gain MOSFET amplifier for the first stage, we can achieve sufficient gain with just four stages.

THE CIRCUIT. The schematic of the low-level amplifier stages is shown in **Figure 4.46**. The first stage is a dual-gate MOSFET amplifier, which is almost identical to the RF amplifier used in the receive converter module. This amplifier has a power gain of 20-25dB. The bandpass filter at the amplifier input was described previously. Note that Gate 1 of the MOSFET is connected to the 'hot' end of L3 and not to the 50 ohm tap near the ground end of the coil. The first amplifier has a tuned output which gives a further reduction in out-of-band spurious signals. The second stage uses a UHF bipolar transistor in the common-emitter configuration. I used a BFY90 transistor for this stage. The BFR91 is more readily available and it also works well in this circuit. The gain of the second stage is about 13dB. The output of the second stage uses a 4:1 impedance matching transformer to give a better match to a 50Ω load. The transformer is a bifilar wound, transmission-line transformer which is wound on a ferrite core. Both stages are biased for Class-A linear operation.

CONSTRUCTION. The low level amplifiers were built using point-to-point wiring on the same sheet of copper clad board as the transmit mixer and band-pass-filter. The same measures that were so effective at curing the instability problems in the receive amplifier were also used in the transmit amplifier. These amplifiers have very substantial gain. To ensure that the amplifiers are stable and well behaved, you must follow good VHF/UHF practice. Component leads should be no longer than necessary. Connections between components should be short and direct. Particular care should be taken with

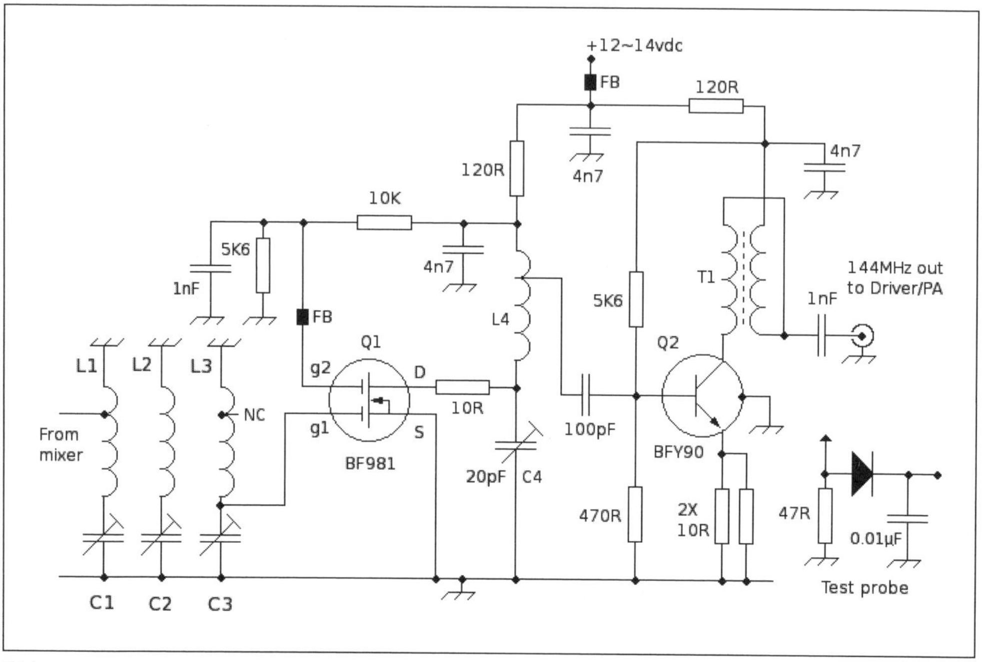

FIGURE 4.46: Two stage low level amplifier. FB=ferrite bead.

PHOTO 4.15: The completed low level amplifier of Figure 4.46.

ground connections and decoupling capacitors. A rectangular section of double-sided PCB laminate was used to make a screen between the MOSFET input and output circuits. Two small slots were cut at the bottom of the screen, one for the drain terminal of the MOSFET and the other for the Gate 2 voltage supply. A small ferrite bead was placed on the Gate 2 lead and on the wire providing the DC power supply to both stages. If you use a BFY90 transistor for the second amplifier, the transistor lead which is connected to the metal body of the transistor should be soldered to the PCB ground foil. I also soldered the transistor body (metal can) to the PCB. This makes a very effective heatsink. Be careful not to apply excessive heat to the transistor. If the copper foil and the transistor body are properly cleaned and pre-tinned, you should be able to make the solder joint in a few seconds. The construction of L4 is exactly the same as L1 and L3. This inductor has 6 turns of 1mm diameter bare copper wire wound on a 5mm former with a tap at 1.25 turns from the DC supply end of the coil. C4 is a 20pF trimmer capacitor of the same type used in the BPF. All fixed capacitors are standard disc ceramic types. All resistors are 0.25W carbon film or similar. T1 is 4 turns made from a twisted pair of insulated wire, bifilar wound on a type 43 ferrite toroid. The assembled circuit is shown in **Photo 4.15**. You might be able to see the 47Ω resistor and germanium diode that were used to measure the output power of the amplifier at the right hand side of the photo.

TESTING. The procedure for testing the amplifier stages is much the same as the procedure that was used for testing the mixer and band-pass filter. Connect a 12-14 DC power supply to the amplifier stages and measure the voltage drop across the 120Ω resistors in the collector and drain circuits of the two stages. Ohm's law tells us that the current through each amplifier is: I=V/R. You should find that each amplifier draws 15-20mA from the supply. The local oscillator module and 10m transmitter should be connected in the same manner as described for the previous test. If you have a 2m receiver, you should repeat last month's filter alignment procedure. C4 should also be adjusted for maximum output at the centre of the band. I used a 47Ω resistor as a dummy load for the output of the amplifier and a simple diode probe to measure the peak output voltage. The diode is a germanium point-contact type 1N34A, OA91 or similar. The voltage drop across a point contact diode varies with the current through the diode, the diode temperature and the characteristics of the individual diode. At very low current levels, the forward voltage drop can be as low as 0.1V. At higher current levels, the voltage drop will be much greater, typically about 1V at 5mA. You should allow for the voltage drop across the diode when you are measuring the output voltage. We will assume that the forward voltage drop of the diode is 0.2V.

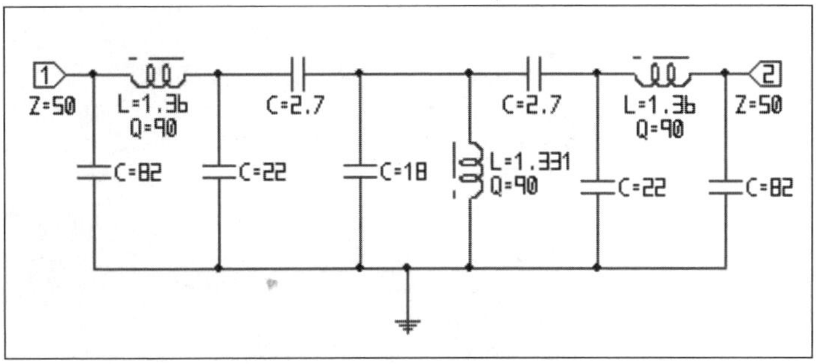

FIGURE 4.47: 28MHz bandpass filter to reduce spurii.

At normal SSB drive levels, the measured peak output is 2V. This becomes 2.2V when the voltage drop of the diode is accounted for. The output power is V^2/R where V^2 is the RMS voltage squared and R is the load resistance. $(2.2 \times 0.707)^2/47 = 51$mW. This might not seem like a lot of power, but it is a substantial increase from 20 microwatts. My record distance for this power level on 2m is more than 300km. Now that we have a couple of volts of RF output, it will be possible to align the BPF properly. Tune C1, C2, C3 and C4 for maximum output at the centre of the band. Check the power output at the upper and lower band edges to make sure that the filter is correctly aligned.

28MHz BAND-PASS-FILTER. The spectral purity of the output signal will depend on the quality of the 10m drive signal. To put it bluntly, if you put rubbish in one end, you will get amplified rubbish out the other end. The signal from the 10m exciter must be absolutely clean. It is fairly easy to generate a very high quality SSB or CW signal when the maximum power requirement is in the 100 to 200mW range. This signal must be free of harmonics, IF image or any other out-of-band spurii. It is a good idea to use a low-pass filter or preferably a band-pass filter between the 10m transmitter/transceiver and the transverter. The band-pass filter shown in **Figure 4.47** will provide useful attenuation of out of band signals. The insertion loss of the filter is just over 1dB. A power vs frequency plot of the filter response is shown in **Figure 4.48**. All of the capacitors are ceramic disc type. The three inductors are 9-10 turns wound on a slug-tuned former. The nominal inductance is 1.4μH. The tuning slug will allow the inductance to be varied over a significant range. When the slug is removed completely, the inductance will be reduced to less than 0.5μH. If suitable cores are not available, you might prefer to use fixed inductors and replace the 18pF and 22pF fixed capacitors with 60pF variable trimmer capacitors. 18 turns on a T50-6 (yellow) powdered-iron core should be about right. **Photo 4.16** shows the assembled 10m BPF.

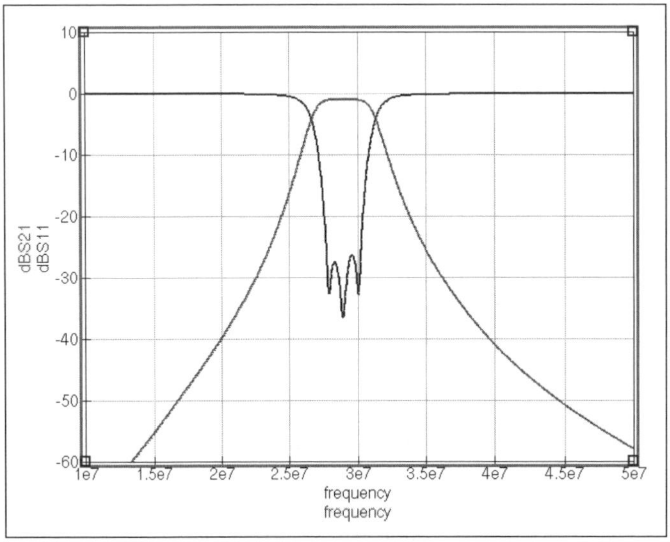

FIGURE 4.48: Bandpass filter characteristics.

PHOTO 4.16: The prototype bandpass filter of Figure 4.47.

RF POWER MODULES. At this stage of the project, you have to make an important decision. Should you build the driver and PA stages from discrete transistors or should you use a monolithic RF power module? VHF power transistors are quite expensive and very easily damaged. Many transverter builders prefer to use a power module which contains the driver, PA and I/O coupling networks which provide a good match to 50Ω. These modules are quite literally a 'black box' solution to the problem. Some of the Mitsubishi range of MOSFET RF power modules are designed for an input power of 50mW. This makes them a perfect match for the low-level amplifier stages of our transverter. The cost of these modules is roughly £1-£2 per watt which makes them quite affordable. Mitsubishi RF modules are available from specialist VHF component suppliers like GH Engineering. The RAxxH1317M **[47]** range of modules are particularly interesting. The Mitsubishi RF module data-sheets have detailed information and application notes. If you intend to use one of these power modules, you can skip the driver and PA sections.

DRIVER STAGE. The driver stage is a very simple common-emitter amplifier. This amplifier can deliver a power output of 500mW into a 50Ω load. I used a 2SC2329 transistor for this stage. This is a VHF transistor in at TO-39 metal can. Unlike most other VHF/UHF transistors, the metal body of the transistor is not connected to the collector terminal of the transistor. This means that there is no need to insulate the transistor from ground. In fact, I did just the opposite and soldered the metal can to the copper ground plane of the PCB. This makes an excellent heatsink for the transistor. You can use a 2N4427 transistor instead, but you will have to use a clip-on heatsink. The collector of the 2N4427 is connected to the case of the transistor, so you will need to isolate the transistor and heatsink from ground. Because the amplifier is biased for class-A operation, the smallest type of clip-on heatsink will not provide adequate cooling. A larger 30°C/Watt or better heatsink will be required.

DRIVER CONSTRUCTION. The driver and PA stages were built as a single module using the same point-to-point construction methods that were employed for the earlier stages. The driver is quite simple to build because there are no matching networks or filters in this stage. **Photo 4.17** shows the driver and PA

PHOTO 4.17: The completed driver and PA circuit.

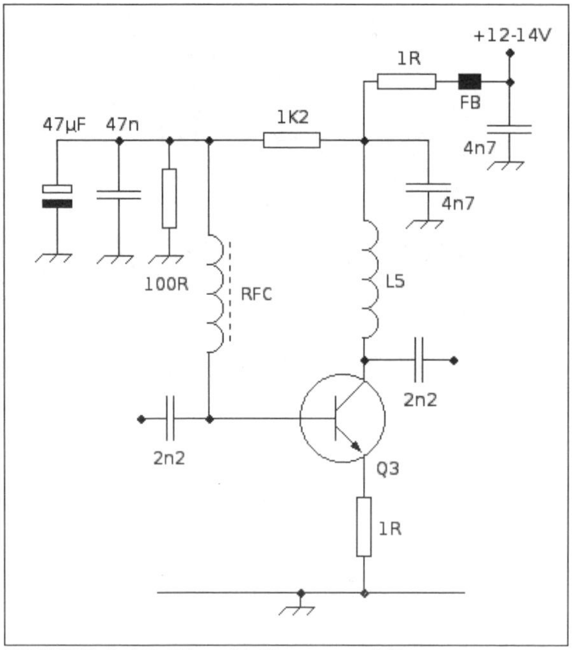

FIGURE 4.49: Circuit diagram of the driver amplifier.

stages. **Figure 4.49** shows the schematic of the driver amplifier. The 2SC2329 transistor (Q3) is soldered to the PCB foil. As with the BFY90 in the low-level stages, you should take care not to apply too much heat to the transistor. The RFC (radio frequency choke) which supplies base bias for Q3 is made from 3 turns of enamelled copper wire wound on a small ferrite bead. The diameter of the wire is not too critical. The wire I used has a diameter of about 0.3mm. L5 is 3.5 turns of 0.7mm enamelled copper wire. The inside diameter of the coil is 5mm. Once again, a 5mm drill bit was used as a coil former. The leads of the 1 ohm resistor should be kept as short and straight as possible.

PA STAGE. The final stage of the transverter is the PA (power amplifier). The amplifier is loosely based on the 28MHz PA design constructed previously. I used a 2SC2539 **[48]** power transistor for Q4. This is an emitter-ballasted VHF power transistor that is rated for 10-14 watts output at frequencies up to 175MHz. The transistor is biased for class-AB linear operation.

PA CONSTRUCTION. I used a power drill and nibbling tool to cut a 10mm x 27mm rectangular slot out of the double-sided PCB. This allows the transistor to be mounted directly on the heatsink. A thin film of heatsink compound should be used between the transistor's mounting flange and the heatsink. The transistor is bolted to the heatsink with a pair of M3 screws, nuts and flat washers. To avoid putting unnecessary strain on the emitter leads, they should not be soldered to the copper ground foil until the PA transistor and the driver/PA board have been properly mounted. You should use a reasonably large heatsink like the 1.8°C/W Maplin heatsink used for our recent HF PA project. Don't depend on the metal case of the transverter for additional heatsinking. Heating of the entire enclosure is likely to cause excessive frequency drift of the local oscillator.

The two 65pF trimmer capacitors in the input matching network are standard (yellow) film dielectric types. L6 is 2 turns of 0.7mm enamelled copper wire. The inside diameter of L6 is 5mm and the turn spacing is equal to the wire diameter. L7 and L8 are made from 1mm bare copper wire (TV co-ax centre core) with internal diameter (ID) of 5mm. L7 is 3 turns, L8 is 4 turns. The turn spacing for both coils is the same as the wire diameter. The 30pF trimmer capacitors in the output matching network are air-spaced or high-quality mica/ceramic compression types. Don't be tempted to use ordinary film dielectric or miniature ceramic trimmers. The 22pF fixed capacitor is a metal clad mica or a high quality silvered mica type. You would probably get away with using a larger high voltage disc ceramic, but you should not use miniature ceramic capacitors for this application. The RFC in the base bias circuit is 3 turns of enamelled wire on a small ferrite bead. The IN4001 diode in the base bias circuit controls the bias voltage. It should be mounted in close thermal contact with the PA transistor. The 180 ohm resistor in the bias network is a 2W carbon film type. As this is not part of the RF circuit, you can also use a wirewound resistor for this component.

PA TESTING. Take great care when you are testing the PA stage. See the HF PA project previously for some advice about building and testing RF power amplifiers. You should put a resistor in series with the 13.8V DC positive supply lead while you are checking the DC standing current. A resistor in the 5-10Ω range should limit the maximum current to non-destructive levels. If you don't have a suitable dummy load for testing the PA, you can use three 150Ω 2W carbon film resistors connected in parallel. This load will be able to cope with output levels of 10-12 watts for a short period of time. Put the 50Ω dummy load on the output of the PA and apply 13.8V via the current limiting resistor to the PA stage only. If your power supply doesn't have an accurate current meter, you should put a DC ammeter in series with the supply. I used the 10A range on my digital multimeter for this job. My PA draws 190mA from the DC supply. The DC bias network takes about 73mA which means that the collector current of the transistor is 117mA. If the DC bias conditions seem ok, you should leave the circuit running for a few minutes. Check the temperature of the transistor regularly to make sure that the heatsink is doing its job properly. If all seems well after a few minutes, you should disconnect the power supply and remove the protection resistor from the supply lead.

I used a dummy load and a simple diode probe to measure the peak output voltage of the PA. This arrangement works well and it is probably more accurate than many commercially made RF power meters. You should connect the power supply to the driver and PA stage and apply RF from the low level amplifier to the input of the driver stage. Keep a close eye on the DC current meter to make sure that there is no sign of thermal runaway or RF instability. Use SSB rather than high duty cycle modes like FM or AM for the tests. Set all of the trimmer capacitors in the driver/PA module to about 60-70% of their maximum value. When you speak into the microphone, you will probably find that you can already measure some RF output into the

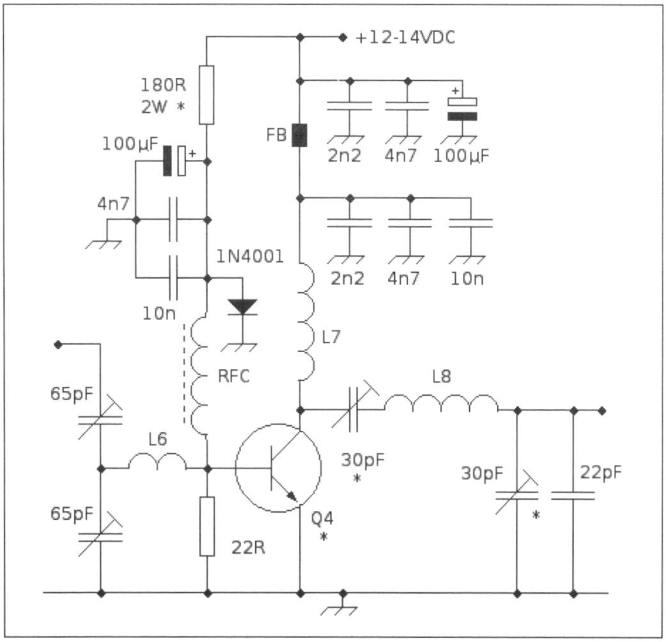

FIGURE 4.50: Circuit diagram of the power amplifier.

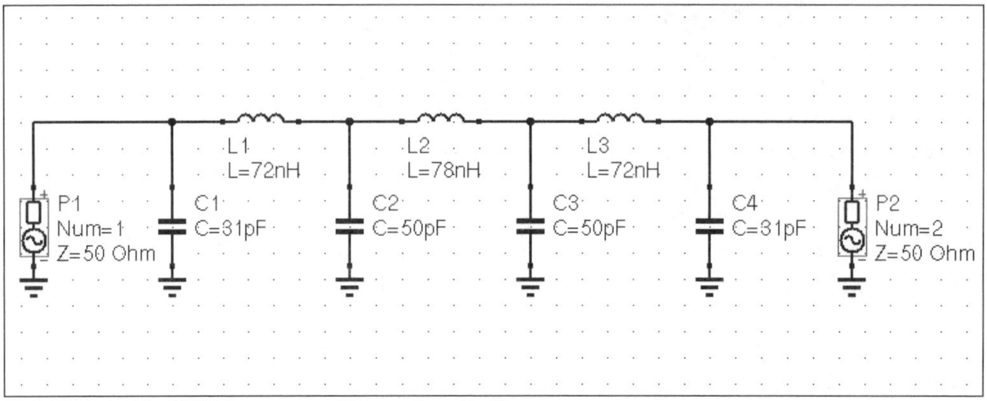

FIGURE 4.51: VHF low pass filter circuit diagram.

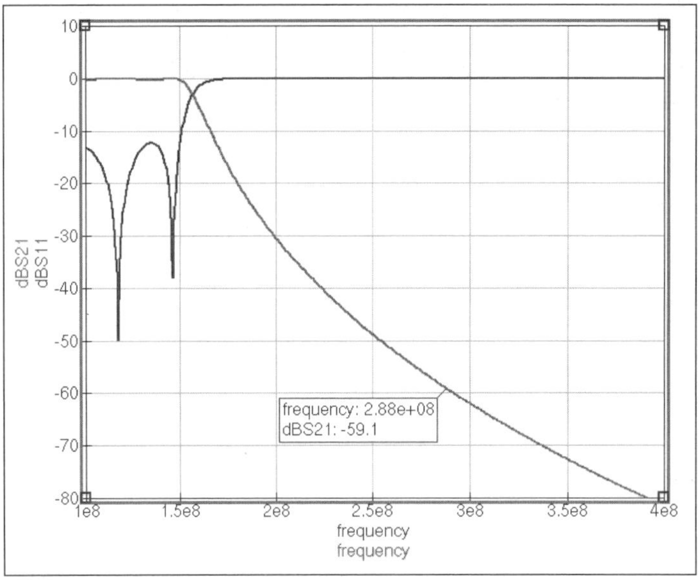

FIGURE 4.52: Low pass filter characteristic.

dummy load. You should adjust the trimmers at the PA input matching network for maximum current draw from the power supply. Next, you should adjust the trimmers in the output network for maximum RF output. There is some interaction between all of these adjustments, so you will need to repeat this procedure several times. I found I could easily get my required output of 5W at moderate 10m drive levels and a saturated output of about 12W with a single tone whistle. I didn't explore the upper limits of the amplifier's capabilities for too long. The Mitsubishi engineers did a very fine job of putting the magic smoke into the PA transistor and I certainly didn't want to do anything that would allow it to escape!

Before we can connect the transmitter output to an aerial for on-air tests, we will need to add a low-pass filter to the attenuation. The low pass filter in **Figure 4.51** will improve the second harmonic attenuation by almost 60dB. **Figure 4.52** shows a plot of the frequency response of this filter. The capacitors are made from a combination of fixed and variable capacitors. L1 and L3 in the LPF are 5 turns, L2 is 5.5 Turns. All coils are 5mm ID and turn spacing is 1mm.

Low Pass Filter and RX/TX Switching

L-C LOW PASS FILTERS. Just about all radio transmitters use some form of low pass filter (LPF) to suppress harmonics and other spurious signals that are at a higher frequency than the transmitter's output frequency. **Figure 4.53A** shows the most basic type of L-C LPF. The filter is made from a pair of reactive elements: the inductor (L) and capacitor (C). The reactance of L is doubled and the reactance of C is halved when the input frequency is increased by 1 octave (doubled). This type of filter is known as an L-network. A typical L-network based LPF will suppress harmonics and other high-frequency spurii at a rate of about 12dB per octave. The L-network has some limitations compared to the other networks in Figure 4.53. Z1 must always be greater than Z2, which means that this type of LPF is not a good choice for use in a system where the input and output impedances are identical. By adding a single reactive element to the L-network, we can transform it into a PI network as shown in **Figure 4.53B** or a T-network as shown in **Figure 4.53C**. The PI and T networks can both be considered as a pair of back-to-back L networks. The PI and T networks that are described here are designed to be terminated at both ends by a 50 Ohm impedance. Such filters are symmetrical and bidirectional. This means that the same filter can be used for transmitting and receiving without the need for any RX/RX switching. These networks are not limited to just three elements; PI and T networks can be expanded to include any number of reactive elements. **Figure 4.54** shows a 5-element PI LPF that is based on the circuit

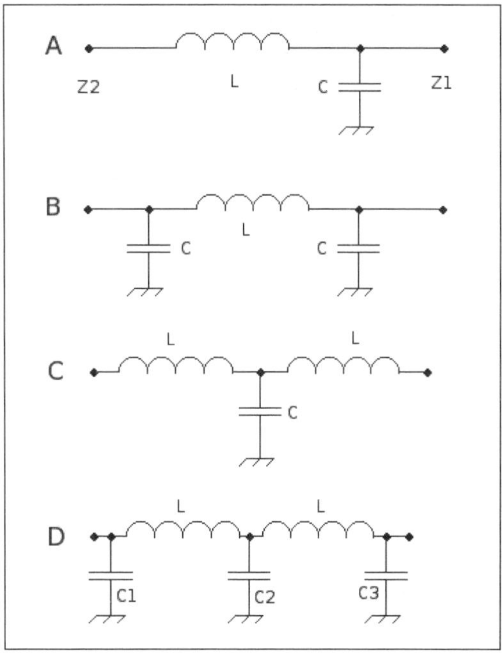

FIGURE 4.53: Low pass filters.
(A) basic LC
(B) PI network
(C) T network;
(D) 5 element PI network based on (C).

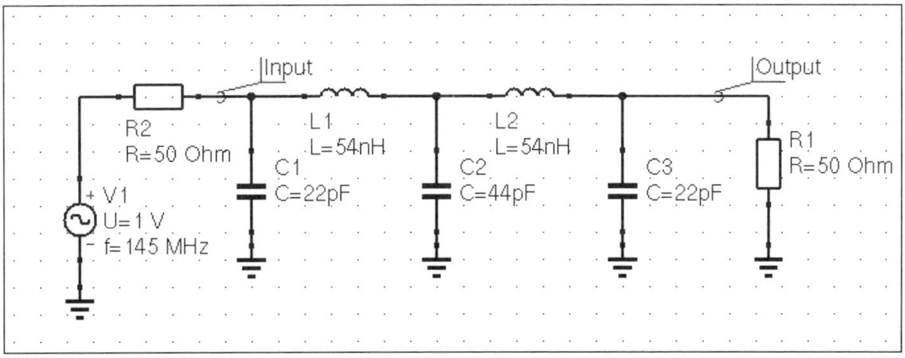

FIGURE 4.54: Practical 5 element PI low pass filter for 2m.

in **Figure 4.53d**. This filter is equivalent to a cascaded pair of 3-element PI networks. Note that the value of C2 is two times the value of C1 and C3.

FILTER TYPES. The PI and T network provide a great deal of flexibility for the circuit designer. They can be used for low and high pass filtering, impedance matching and phase-shift networks. The simple C-L-C type of PI network is often used in valve power amplifiers to match the high output impedance of the valve to a 50Ω load. Despite the apparent simplicity of the PI and T filters, there are many different types of filter that are based on these networks. Each type has its own characteristics. The two most commonly used filter types are known as Butterworth and Chebyshev filters. Butterworth low-pass filters have a maximally flat response within the passband and a gradual roll-off at the cutoff frequency. Chebyshev filters have more passband ripple and a sharper roll-off at the cutoff frequency. Filters with a Chebyshev response employ a trade-off between pass-band ripple and stop-band attenuation.

A SIMPLE LPF DESIGN. We will start by designing a very simple PI type of low pass filter. The filter will have five reactive elements. Since this filter is based on a PI rather than a T network, we will need three capacitors and two inductors. The filter design is based on a three element PI matching network with input and output impedance of 50Ω and a loaded Q of 1. This network is very easy to design because the required value of reactance for all three components is exactly 50Ω. **Figure 4.55** shows the formula that I used to calculate the component values **[49]**. Since the five element filter is effectively a cascaded pair of three element filters, the reactance of C2 is 25 ohms. The frequency response curve of this filter is similar to a Butterworth filter which means that passband ripple will be very low. This type of filter is often referred to as a half-wave harmonic filter because it gives a 180 degree phase shift for signals that are just below the cutoff frequency. **Figure 4.56** shows a QUCS simulation of the filter. The filter input is driven by an AC voltage source at a frequency of 145MHz and the filter output is terminated by a 50Ω resistor. The dotted trace shows the voltage at the filter input and the dashed trace shows the voltage at the filter output. The plot shows a time period of 20 nanoseconds. Note that there is a 180° phase difference between the two signals.

```
XC1 = R1/Q

XC2 = R2 * sqrt((R1/R2)/((Q*Q+1) - (R1/R2)))

XL1 = (Q*R1+R1*R2/XC2)/((Q*Q)+1)

C1 = 1 / (6.28 * F * XC1) microfarads

L1 = XL1 / (6.28*F) microhenries

C2 = 1 / (6.28 * F * XC2) microfarads
```

FIGURE 4.55: The formula used to calculate the component values for the practical filter.

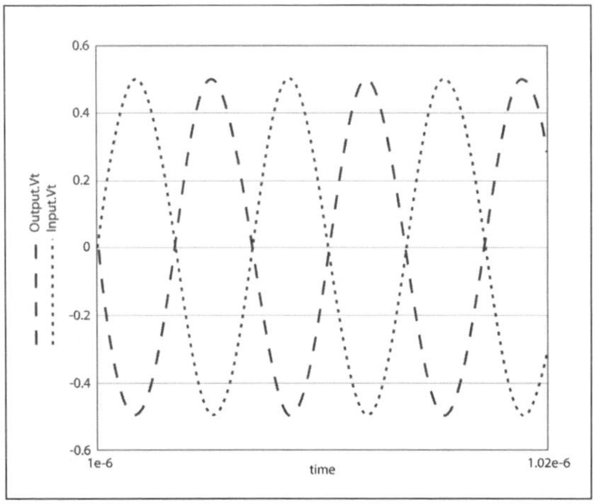

FIGURE 4.56: Simulation output for the filter at 145MHz. Dotted trace = input voltage; Dashed trace = output voltage. Note the 180° phase difference.

The dashed trace (1) in **Figure 4.57** shows the transmission loss through the filter. Two other traces are shown for comparison. The solid trace (2) shows the response of a 5-element Chebyshev filter with a passband ripple of 0.1dB, the dotted trace (3) is for a Chebychev filter with 1dB of passband ripple. Normalised tables for designing Chebyshev filters with various values of passband ripple can be found in the *ARRL Handbook* **[50]**.

A PRACTICAL VHF LPF. This filter was built as part of the 2m transverter project. It can also be used as an add-on low-pass filter for any 2m transmitter or transceiver. Component values and a QUCS plot of the filter response were shown in last month's Homebrew. The measured loss of the filter is less than 0.2dB. The schematic of the filter is shown in **Figure 4.58**. I used the filter design tool in the QUCS **[51]** simulation software to calculate the component values. Dedicated filter design programs like AADE Filter Design **[52]** or on-line web tools http://tinyurl.com/3b2f7s provide a convenient way of fine tuning a design. I started out with the intention of building a Chebyshev filter with a passband ripple of 0.1dB. After a few minutes of tweaking the component values, I found that I could tweak the value of passband ripple so that I could use some readily available capacitors for C1 and C2. The filter is a 7th order Chebyshev PI LPF. The passband ripple is 0.28dB. Since the filter only needs to pass signals between 144MHz and 146MHz, it can easily be tuned so that passband ripple is negligible within the 2m band.

CONSTRUCTION. The LPF was built on two strips of double-sided PCB laminate. **Photo 4.18** shows the layout of the filter. The two strips of PCB are soldered together to form an L shape. The

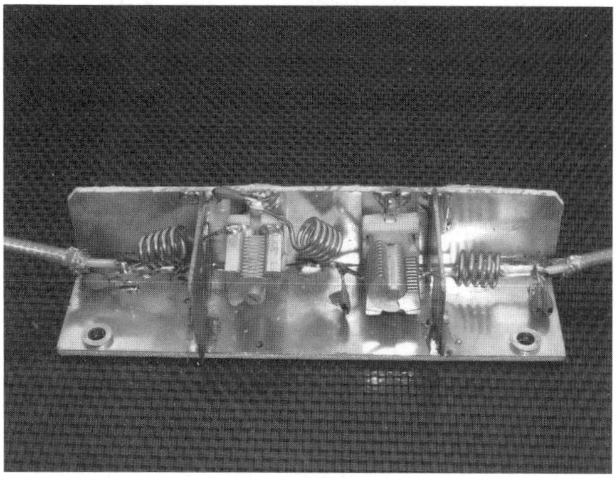

PHOTO 4.18: Construction of the prototype low pass filter of Figure 4.58.

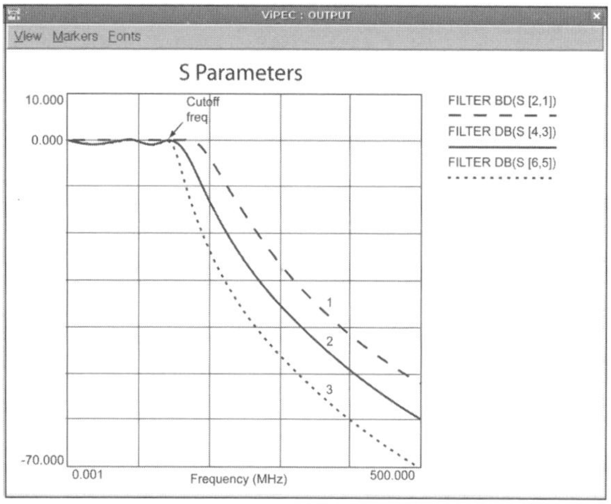

FIGURE 4.57: Transmission loss plots – our filter is dashed, a 0.1dB passband ripple 5 element Chebyshev filter is shown solid while a similar 1dB ripple 5 element Chebyshev is shown as a dotted line.

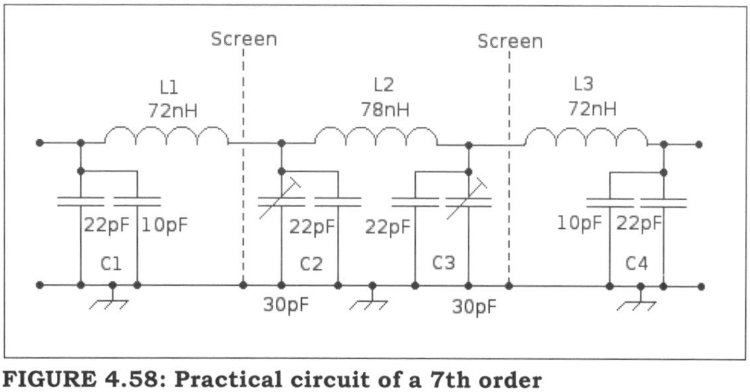

FIGURE 4.58: Practical circuit of a 7th order Chebyshev PI LPF.

copper clad laminate of the vertical section should be soldered on both sides to ensure that it is well grounded and to improve the mechanical strength of the assembly. Two rectangular sections of PCB laminate are used as screens between L1/L2 and L2/L3. A small slot or hole should be cut out of each screen to allow the ends of the inductors to pass through. The vertical PCB section makes a handy mounting point for the two variable capacitors. C1 and C4 are made from a 22pF and a 10pF capacitor connected in parallel. The capacitor leads should be kept as short and straight as possible. One terminal of each capacitor should be soldered directly to the copper ground plane. C2 and C3 are each made from a parallel combination of a fixed 22pF and a variable capacitor with a maximum capacitance of 30-35pF. I used high quality air-spaced capacitors for this job. If air-spaced capacitors are not available, good quality mica or ceramic compression trimmers can be used instead. Miniature film-dielectric or miniature ceramic trimmers are not recommended for this application, particularly if the filter is to be used with any power. L1 and L3 are 5T, 5mm ID. L2 is 5.5T, 5mm ID. 1mm bare copper wire was used for the inductors.

TESTING. I used a home made transmitter with a nominal output of 25W, an SWR meter and a 50 ohm dummy load/power meter to test the LPF. The output power of the transmitter was measured very accurately using a high quality UHF dummy load and a diode probe. The peak RF voltage was just over 49V which indicates that the output power is almost exactly 24W. Next, the LPF was inserted in line between the transmitter and dummy load. C2 and C3 were adjusted for maximum output. During these tests, I noticed that some of the fixed capacitors were getting quite warm! This came as a bit of a surprise because I had gone to a great deal of trouble to find some 22pF silvered-mica capacitors to use in this project. The offending capacitors were replaced with ordinary 100V ceramic disc capacitors. With the new capacitors in place, a peak voltage of 48 volts was measured at the dummy load. The power output is V^2/R or $(48x0.707)^2/50 = 23$ watts. The loss in decibels is 10log(P1/P2) or 10log(23/24) = 0.184dB. This is a surprisingly good result!

For the next test, the SWR meter was placed between the transmitter and the LPF input. The measured SWR was about 1.2:1. A very slight adjustment to C2 and C3 produced a flat 1:1 SWR at 145MHz. Once the filter was correctly aligned at the band centre, the SWR was checked at several frequencies and was found to be better than 1.1:1 across the entire band. The SWR measurements were consistent with the return-loss figures predicted by the QUCS simulation. Unlike the transmit band-pass filter, this is a relatively low-Q filter and it is quite easy to align correctly.

RX/TX SWITCHING. Now that we have built all of the modules for the VHF transverter, we must build a switching system that will automatically route the DC and RF signals to and from the various stages when we switch from receive to transmit. There are several different ways we can approach this problem. Many transceivers use a solid-state switching system to perform the RX/TX switching. This method uses no moving parts and can switch very quickly from receive to transmit and back to receive again. Solid-state electronic switches are immune to the kind of mechanical wear and tear, contact burning and contact oxidisation that can cause failure of electro-mechanical relays. The high speed switching capabilities of solid-state electronic switches makes them particularly suitable for full break-in CW or SSB VOX operation.

There are at least two different types of electronic RX/TX switching. The simple method shown in **Figure 4.59A** uses a pair of diodes to limit the peak voltage at the input of the receive amplifier to a maximum of 0.7V. The value of C is chosen so that it has a high reactance compared to the shunt capacitors in the transmit low-pass filter. This simple voltage clamping method might seem very crude, but it works very well in practice. The circuit in **Figure 4.59B** shows a more sophisticated arrangement which uses a pair of PIN diodes and a quarter-wave resonant circuit to perform the RX/TX switching. Since high power PIN diodes are not readily available in this part of the world, I decided to use a pair of DPDT relays to do the RX/TX switching. Relays are a lot slower than electronic switches but they have a few advantages over solid-state circuits. Relays offer lower losses and better linearity than solid-state electronic switches. They do make some acoustic noise, but they introduce little or no electrical noise into the RF circuit.

CONSTRUCTION. Ideally, we would use a co-axial relay for the 144MHz aerial switch. Since co-axial relays are expensive and difficult to find, we will use a couple of ordinary DPDT (double-pole, double-throw) relays for switching the 144MHz RF, the 28MHz RF and the 13.8VDC signals. The relays I used are Maplin type N31AW. The contacts are rated for 5A at 250V and the coil requires 12VDC at approximately 50mA for each relay. This type of relay can be used as a VHF switch provided that we follow a few simple rules. To keep unwanted radiation from the relay contacts to a minimum, the relay should be mounted horizontally over a conductive ground plane. When using a double-pole relay as a VHF aerial switch, only the lower set of contacts should be used. See **Photo 4.19** for details. RF connections to the relay were made using 50Ω co-ax cable. The braid of the three cables is soldered directly to the copper ground plane. When using ordinary relays for VHF

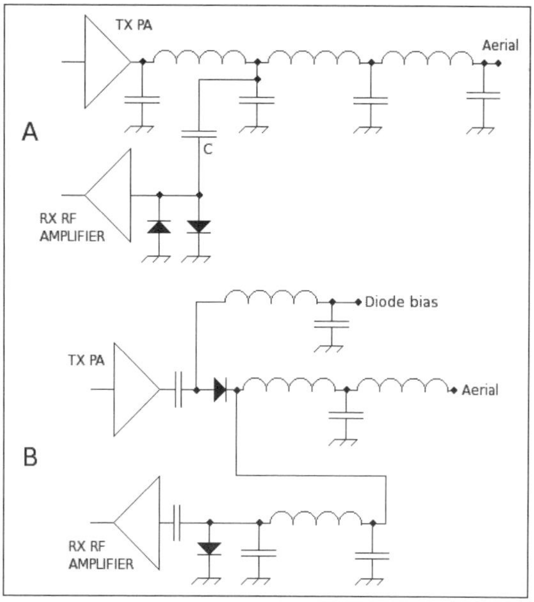

FIGURE 4.59: Solid-state Rx/Tx switching methods.
(A) simple diode clipper
(B) PIN diode method

PHOTO 4.19: How to lay out the RF changeover relay.

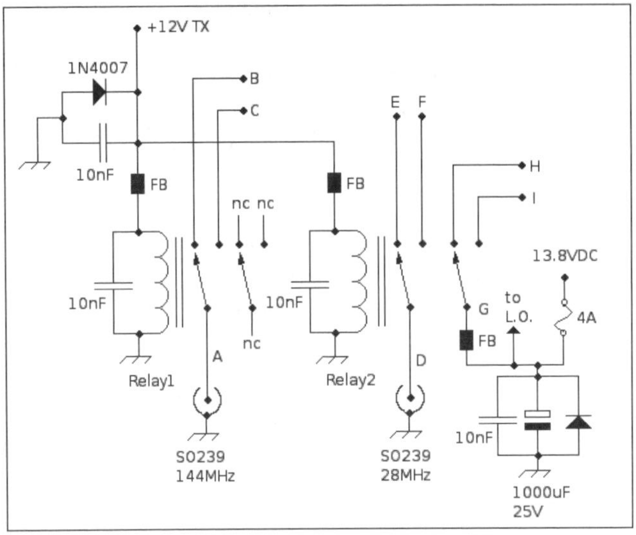

FIGURE 4.60: Circuit diagram of the relay switching arrangement.

switching, you should not assume that there will be very high isolation between the relay contacts and the DC coil wiring. One end of the coil should be soldered directly to ground and the positive DC terminal should be bypassed with a 10nF decoupling capacitor. A ferrite bead and a second capacitor is used to make absolutely sure that none of the VHF output signal finds it way back to earlier stages via the DC wiring for the relay.

The second relay is used for switching the 28MHz IF signal and the DC supply to the transmit and receive stages. Because of the lower frequencies and RF power levels, this circuit is much less critical than the aerial relay. The DC coil supply to this relay is decoupled by a ceramic disc capacitor and a ferrite bead. A 1A rectifier diode is used to absorb voltage transients caused by back-EMF from the relay coils. **Figure 4.60** shows the schematic of the relay switching circuit. The DC power input circuit is protected by a 4A fuse and a IN5408 rectifier diode which offers some protection against accidental cross-wiring of the DC supply leads.

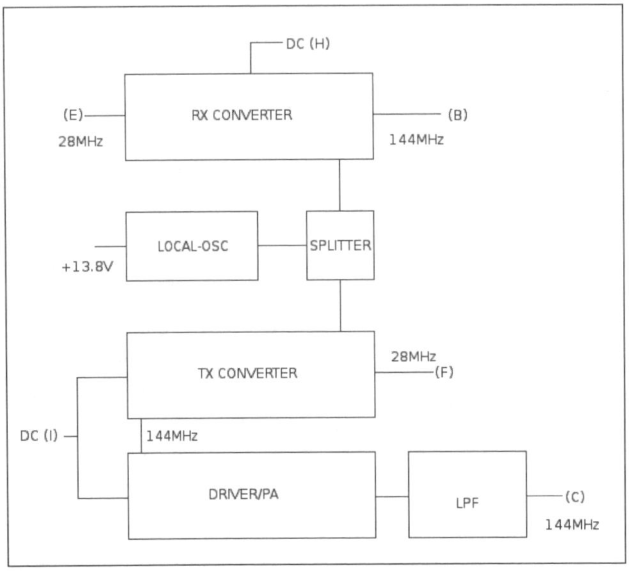

FIGURE 4.61: Block diagram of the complete transverter.

Figure 4.61 shows a block diagram of the completed transverter. Each module should be properly screened and grounded. The LPF module and aerial changeover relay should be mounted close to the 2m aerial connector at the back of the enclosure. The RX/TX relays are operated by a 12V control line from the 28MHz transceiver. This is by far the best method of transmit/receive switching. The 12V control line should be well decoupled at the point where it enters the transverter enclosure. a 1000pF bolt-in feedthrough capacitor is ideal for this job.

TESTING. Once all of the modules have been installed in the enclosure, you will probably find that you will have to repeat the alignment procedure for some of the modules. Check everything carefully, make sure that the RX/TX relays are operating correctly before you apply power to the transmit mixer. My 10m transceiver has a maximum output of just 200mW. This is probably not enough to destroy the receive mixer in the event of an RX/TX switching failure. If your transceiver doesn't have a transverter output, you should be very careful at this stage. Make sure that your 28MHz transmit drive is going to the attenuator in the transmit converter stage and NOT to the output of the receive mixer! Do a few tests into a dummy-load to make sure that the signal is clean and stable before you put the transverter on the air. My first two-way contacts with the transverter were during the 144MHz May Contest on 19 May. Although tropo conditions were quite poor, I did manage to work several stations including G0VHF/P (JO01) at 663km, G5LK/P (IO91), G2KF/P (IO70) and G2BQY/P (IO81).

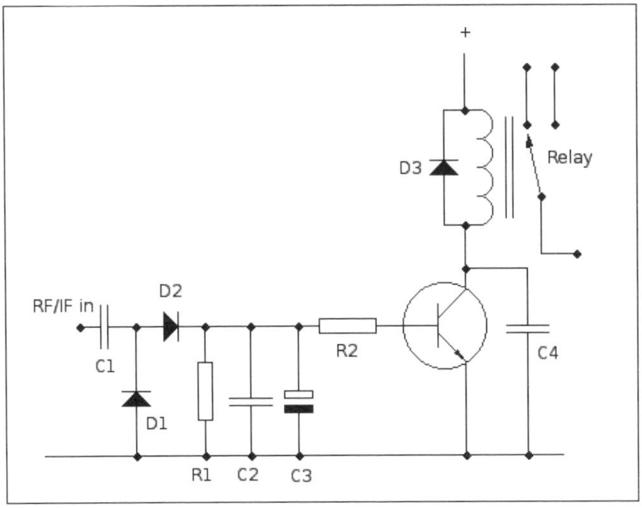

FIGURE 4.62: Typical RF sensing switch which can be used if desired.

I have a great dislike for RF sensing RX/TX switching systems. A hard-wired switching circuit will always do a much better job. However, if you are using a transceiver without any provision for transverter DC switching, or a converted CB, you may prefer to use an RF sensing circuit. The circuit in **Figure 4.62** shows a typical RF sensing switch. The operation of the circuit is quite simple. Some of the 28MHz drive signal is sampled via C1 and fed to a voltage doubling rectifier. This charges up C3 and switches on the transistor. The value of C3 is chosen to give a short delay during speech pauses or between Morse characters (semi-break-in) before the circuit drops back into receive mode.

References

[38] http://tinyurl.com/jsnr6
[39] *Solid State Design for the Radio Amateur*. Hayward/DeMaw. ARRL
[40] http://tinyurl.com/rygb8
[41] Philips BRY39 Programmable unijunction transistor / Silicon controlled switch – data sheet. 1997
[42] Bourns. TIC246 series data-sheet. www.bourns.com

[43] http://tinyurl.com/yqzwcf
[44] http://tinyurl.com/yvemkg
[45] An N-way Hybrid Power Divider Ernest J. Wilkinson. IRE 1960
[46] VHF/UHF Manual c6p6 G. R. Jessop G6JP. Fourth edition 1985
[47] Mitsubishi power modules. http://tinyurl.com/2s8qjp
[48] Mitsubishi Electric 2SC2539 datasheet. Nov. 97
[49] ARRL electronics data book. 1976
[50] ARRL Handbook. 1993. Chapter 2
[51] http://qucs.sourceforge.net/
[52] http://www.aade.com/

5

Aerials

HISTORY. Electro-magnetism was discovered over two centuries ago when Italian amateur scientist Gian Domenico Romagnosi found that an electric current from a simple battery could deflect the needle of a compass **[53]**. Unfortunately, this discovery was ignored by the scientific community at the time. Later experiments by the Danish physicist H C Ørsted in 1820 confirmed the link between electricity and magnetism. During the 1820s the English scientist Michael Faraday carried out experiments in electro-magnetism that would eventually lead to the development of the electric motor. In the mid 1800s, the Scottish physicist J C Maxwell developed a set of equations that describe the behaviour of electro-magnetic fields. Maxwell's equations predicted that electric and magnetic fields would propagate through free space at the speed of light.

Experiments by Heinrich Hertz in 1887-88 proved that many of Maxwell's theories were correct. Hertz used a crude spark transmitter that was connected to a simple dipole transmitting aerial. The receiving aerial was a loop that was broken by a small spark-gap. When the receive aerial was exposed to the electro-magnetic waves from the transmitter, the RF current induced in the loop caused a spark that was visible in a darkened room. Using this simple apparatus, Hertz was able to prove that radio waves travelled at the speed of light – as Maxwell had predicted. He was also able to show several other similarities between radio signals and light. His experiments proved that radio waves could be reflected, refracted and focused into a beam by using a concave reflector. This supported Maxwell's theory that light itself in an electro-magnetic phenomenon. It would appear that Hertz did not fully appreciate the significance of his findings. He is reported to have said to his students: "*It's of no use whatsoever*", "*this is just an experiment that proves Maestro Maxwell was right - we just have these mysterious electromagnetic waves that we cannot see with the naked eye. But they are there*". Today, the standard unit for measuring frequency is named after Hertz. The development of practical radio transmitters would be left to other experimenters and inventors like Bose, Tesla, Popov and Guglielmo Marconi **[54]**.

TERMINOLOGY. The British English word 'aerial' has the same meaning as the American English word 'antenna'. The two words can be used interchangeably. You will often see similar or even absolutely identical aerials referred to by several different names. Even a very simple aerial like a centre-fed wire doublet might be referred to as a dipole, doublet or double-Zepp. Depending on the electrical length of our simple doublet, it might be described as a half-wave dipole, EDZ (extended-double-Zepp), G5RV and so on… The same is true of end-fed aerials, which are often described as end-fed wire, random wire, long wire, Marconi end-fed etc.

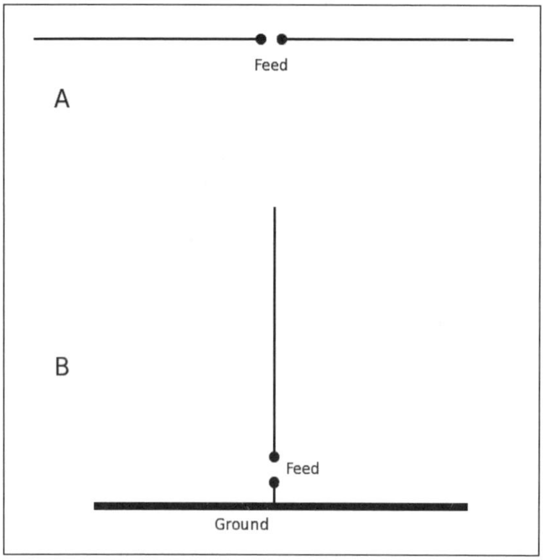

FIGURE 5.1: The basic types of aerial.
(A) dipole, or Hertzian
(B) End-fed, or Marconi

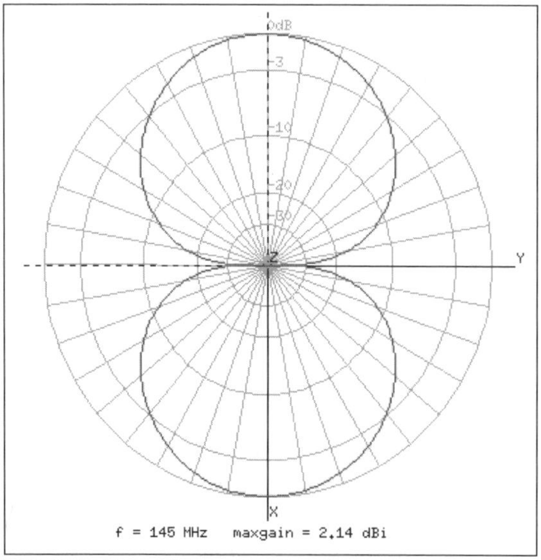

FIGURE 5.2: Theoretical radiation pattern of a half-wave dipole in free space.

TYPES. The most basic type of radio aerial consists of a pair of conductors which are arranged as shown in **Figure 5.1A**. The two wires run in opposite directions and are usually of equal length. The feed point is located at the centre of the dipole. This is the type of aerial that Hertz used as part of his transmitter. Aerials that do not require an earth connection are known as a Hertz aerial or Hertz antenna. Loop aerials like Hertz' receiving loop, quad loops and other similar types are also categorised as Hertz or Hertzian aerials.

The other basic type of aerial usually consists of a single conductor which acts as the radiator. The second terminal of the feed system is connected to ground (earth). The ground can consist of buried conductors like an earth rod, a continuous ground-plane like the roof of a motor vehicle or a system of radial wires. This type of aerial is commonly known as a Marconi aerial. This category includes end-fed verticals (which are often referred to as ground-plane aerials or monopoles, especially if the radiating element has an electrical length of 1/4 wavelength), Inverted-L, Marconi-T and off-centre fed aerials with a single wire feed like the Windom. **Figure 5.1B** shows a ground mounted vertical monopole.

GAIN. Aerial gain is usually measured in decibels relative to some known reference. The two most commonly used references are an 'isotropic radiator' and a dipole. An isotropic radiator is a purely theoretical device which radiates equally well in all directions. Gain relative to a dipole is expressed as dBd and gain relative to an isotropic radiator is dBi. The free-space gain of a perfect lossless dipole is 0dBd or 2.15dBi.

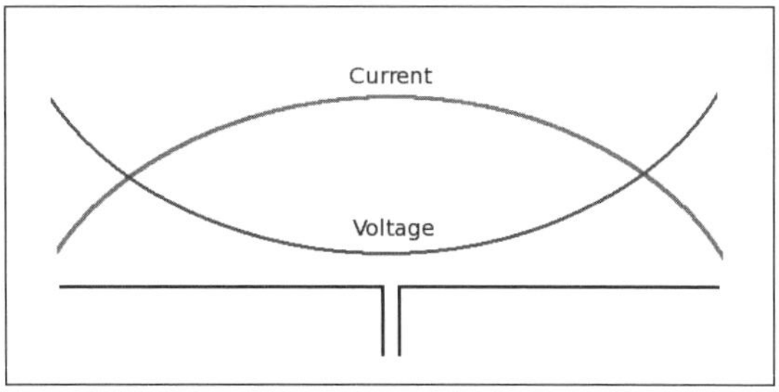

FIGURE 5.3: Voltage and current distribution along a half-wave dipole.

THE DIPOLE. The simple half-wave dipole is one of the most popular homebrew aerials. Dipoles are very easy to make from readily available materials. Obviously, a passive device like a dipole can't give a real power gain. It can't possibly radiate more power than is available at the feed point. However, it can show an apparent gain in a favoured direction if this gain is accompanied by a relative loss in another direction. **Figure 5.2** shows the radiation pattern of a half-wave dipole in free space (if such a place were to exist). The dipole has a gain of 2.15dBi in a direction which is broadside to the plane of the dipole. There are deep nulls in the radiation pattern off the ends on the wires. The feed point impedance of a resonant centre fed half-wave dipole is approximately 73Ω. **Figure 5.3** shows the voltage and current distribution along the length of a half-wave dipole. The maximum current is at the centre and the maximum voltage is at the ends. Ideally, the dipole would be connected to the transmitter or receiver by a length of transmission line which has a characteristic impedance which is equal to the dipole feed point impedance. 75Ω balanced line is well suited to this task. 75Ω co-axial cable is also suitable if a 1:1 balun is used at the dipole feed point. Since neither terminal at the feed point of the dipole is connected to ground, the RF voltage and current at the dipole feed point will tend to be balanced with respect to ground. The balun (balanced to unbalanced) transformer is a device which allows the balanced feed point of the dipole to be connected to an unbalanced co-axial feed line without introducing any unwanted RF (common mode) currents on the feed line. The balun can be a simple transmission line transformer wound on a ferrite core or it can be made from a length of co-ax line. The simplest form of choke balun can be formed by making a multi-turn loop at the aerial end of the co-ax feed line.

VERTICAL MONOPOLE. The vertical monopole is the most common type of mobile aerial. Most mobile whips use the metal body of the vehicle as a ground plane and a flexible whip as the radiating element. VHF/UHF whips are usually at least ¼ wavelength long. HF mobile whips are usually tuned to resonance by a series inductance. Full size, ground mounted verticals are very popular with LF band DX operators because they tend to produce good low angle radiation. Horizontal aerials which are mounted a small fraction of a wavelength above ground are good high angle radiators. Such aerials can be very effective for working local stations at distances up to a few hundred km, but they are relatively poor for working DX stations. Unlike the balanced dipole above, vertical monopoles are unbalanced and will require a good earth connection if they are to work efficiently. Theoretical models of ground mounted verticals often assume that the ground is an infinitely large ground

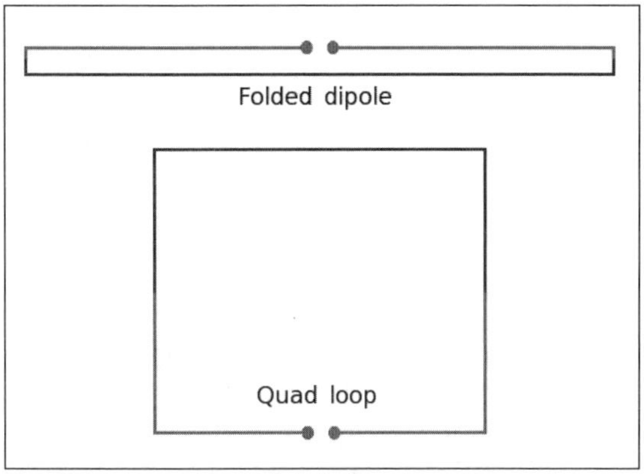

FIGURE 5.4: The folded dipole and quad loop can be simplified into pairs of joined dipoles.

plane made from a perfect conductor. A 1/4 wavelength vertical monopole can be considered to be equivalent to one half of a dipole. The ground connection acts as the other half of the dipole. If we could find a perfect lossless ground plane, the feed point impedance of our resonant monopole would be exactly half the impedance of a dipole or about 36.5Ω. As there will always be some losses in the ground system, the feed point impedance will tend to be higher than this value.

OTHER TYPES OF ANTENNA. There are many different types of radio aerial, but all of them can be considered as dipoles or collections of dipoles. A full-wavelength loop like the quad loop or the diamond quad is really just two dipoles that are joined at the ends. The folded dipole as shown in **Figure 5.4** is a pair of dipoles which are connected in parallel. Yagi arrays are made up from multiple dipoles arranged on a common boom. Other 'gain' aerials like multi element quad beams, beams made from arrays of circular loops, Rhombics, V-beams, log-periodic arrays, long wires and so on are all derived from the dipole or monopole.

TRANSMISSION LINE. Except for a few rare situations where the transmitter or receiver is located directly at the feed point of the aerial, it will be necessary to use some form of transmission line to connect the transmitter/receiver to the aerial. (In this context, 'transmission' means 'conveying from A to B' rather than anything necessarily to do with a transmitter). The two most common types of feed line are co-axial cable and balanced twin feeder. Co-ax with a characteristic impedance of 50 or 75Ω is readily available. Balanced twin feed line is available as 75Ω, 300Ω or 450Ω 'flat twin'. 600Ω open wire balanced line is very to make from insulated wire and plastic insulators.

A MULTI-BAND HF DOUBLET. This month's project is a multi-band HF aerial. There is always some ambiguity about the exact meaning of the word dipole. This word is usually used to refer to a centre fed aerial which has an electrical length of approximately ½ wavelength (180 electrical degrees) or to an electrically short dipole which has a physical length which is significantly less than a half wavelength. In the case of a wire doublet with a total length of one wavelength, would it be correct to call this a dipole or would it be more appropriate to refer to it as a co-linear array consisting of two dipoles? There is even more scope for confusion when the aerial is designed for multi band use so that it might act as a half-wave dipole on one frequency band and not on others. To remove this ambiguity, centre fed wire aerials that are electrically longer than ½ wavelength are often referred to as doublets.

The HF doublet is to be used on all of the nine HF bands. I use a 600Ω open wire line and a balanced tuner so that the system can be tuned for operation on every band. The total length of the doublet is 38 metres which is a half wavelength on the 80m band. It works as a collinear array of two dipoles on the 40m band and as a 2 x 5/8 collinear array on 30m. On the higher frequency bands, the radiation pattern tends to split up into multiple lobes. The horizontal radiation pattern on 20m resembles a four leaf clover. On 17, 15 and 10m the radiation pattern is more complicated. I have had quite a few contacts on the 6m band using this aerial and I have even worked a few distant stations on 2m with it.

The doublet is arranged in an inverted-V configuration. The centre insulator is attached to a fibreglass pole which is mounted on top of a steel pole, which is fixed to the gable end of our two-storey house using steel stand-off brackets. The highest point of the doublet is more than 12m above ground. The ends of the doublet slope downwards to a pair of supports. The support in the front garden is a length of 3 x 2 inch timber about 3.7m high. The support in the back garden is a steel scaffolding pole which is about 6.1m high. The garden is not quite long enough to accommodate the total length of the doublet so the end of one side of the doublet is bent into a 'dog-leg' to make it fit into the space available. This has no significant effect on the performance of the doublet. **Photo 5.1** shows the centre of the HF doublet and a 2m Yagi.

PHOTO 5.1: The centre of my doublet can just be seen below the 2m Yagi.

CONSTRUCTION. Construction of the doublet is very straightforward. The only hard work involved it the erection of the three supports. The doublet and the feed line are made from 2.5mm standard earth wire. This wire is very cheap and readily available from just about any electrical supplier. When I made my first HF doublet, I measured and cut the two halves of the doublet and the two sides of the balanced line separately. The feed line was then soldered to the doublet or even worse, joined with a 'chocolate block' connector. After several feed line breakages, I realised that this was the wrong approach. The feed line and the doublet should be made from two continuous lengths of wire.

The feed line insulators were made from 170mm lengths of 16mm electrical conduit. This conduit is available from DIY shops such as B&Q and most electrical suppliers. Each length of conduit was drilled at 10mm from the end, giving a wire spacing of 150mm. **Photo 5.2** shows the insulators after drilling. The ends of the two feed line wires should be tied to a solid anchor so that the wires can be pulled straight and the insulators can be pulled into place. Use as many insulators as

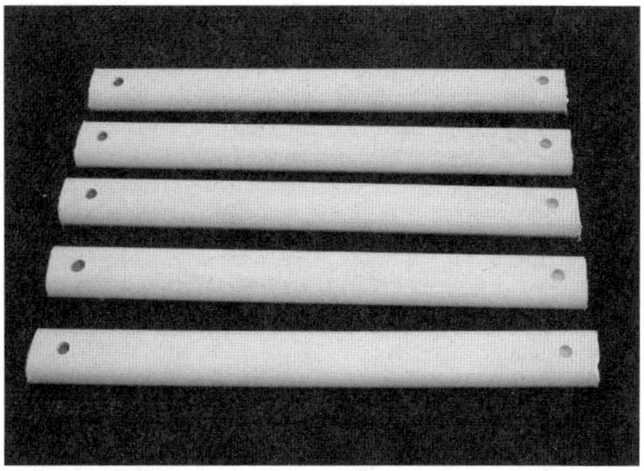

PHOTO 5.2: The home-made spreaders after drilling. Each spreader is about 170mm long and drilled 10mm from the end, to give a wire spacing of 150mm.

are required to keep the wires straight and parallel. I use a spacing of 1m between each insulator on the straight run from the roof of my shack to the feed point. I use closer spacing for the feed line that is inside the shack. **Photo 5.3** shows my feed line making jig, which is otherwise known as the front garden railings. When the insulators are spaced correctly, each one should be fixed in place with two short lengths of 1mm copper wire. As usual, I used TV co-ax centre conductor for this job. A pair of pliers was used to twist the wire around the PVC insulation of the feed line wire. See **Photo 5.4** for details.

I used a 200mm length of strong plastic pipe as the centre insulator for the doublet. I didn't use the 16mm conduit that was used for the feed line insulators/spreaders for the doublet centre insulator because I wasn't sure that it would be strong enough for this job. The plastic pipe was drilled at 25mm from each end and the wires were passed through the holes. The insulated wire was bent through 90° at the top side of the insulator and fixed firmly to the ends of the insulator with vinyl tape and/or plastic cable ties. This makes a very strong support for the centre of the doublet. I have been using this method for more than 20 years and so far I haven't seen a single failure. When the doublet is raised into position, you should try to arrange the open-wire feed line so that it is not too close to the support mast or any other conductors. I have an outdoor shack so it is relatively easy to bring in the feed line to the tuner. For a more typical installation in an indoor shack, it is quite acceptable to use a short length of 300Ω slotted ribbon between the tuner unit and the 600Ω line outside. Slotted ribbon can usually be squeezed through a 10-11mm hole in a wall, door or window frame. This is the same size hole that would be used for RG8 or UR67 co-ax.

The ends of the doublet should be attached to the end supports using suitable insulators. I have never felt the need to use special insulators for this purpose. A short length of polypropylene rope makes a very good insulator. If you insist on having a 'proper' insulator, you can use electric fence insulators from your nearest farmer's co-op store.

PHOTO 5.3: Making ladder line by the metre using the garden wall as a handy anchor point.

PHOTO 5.4: Use short lengths of 1mm copper wire to fix the spreaders to the cable. Note that the wire passes through the spreader.

I have used various types of tuning unit to match the balanced feeder to the 50Ω output of my transmitter. A proper balanced tuner with a link-coupled output is probably the best option. My current setup uses a simple T-match tuner and a ferrite cored balun to handle the unbalanced to balanced transformation. This works well for me. The balun stays cool under all operating conditions. RF current in the earth lead of the tuner is minimal, which indicates that the system is well balanced. It is quite easy to tune for a perfect 1:1 match on all HF bands. The dipole is a bit too short to work efficiently on 160m so I usually disconnect one side of the feed line from the back of the tuner and load up just one side of the doublet as an inverted-L for 160. This system works well enough to allow me to make the occasional trans-Atlantic QSO on 160m.

Directional Aerials

OMNIDIRECTIONAL? All aerials are directional to some extent. Even those aerials that are described as omni-directional like the vertical ground plane and various other forms of end-fed and centre-fed verticals are only omni-directional in the horizontal plane. Such aerials show gain towards the horizon and a relative loss in other directions. In practice, the maximum gain will be at some point above the horizon because of the effects of reflected signals from the ground.

Horizontal single element half wavelength dipoles and full wavelength quad loops are bi-directional. These aerials have the familiar figure-of-eight free space radiation pattern where the maximum gain is broadside to the plane of the aerial with deep nulls off the ends. There is really no such place as 'free space' and in the real world, the radiation pattern of all practical aerials is influenced by the presence of the small blue-green planet under our feet and other objects within a few wavelengths of the aerial. When a horizontal aerial is placed relatively close to the ground, the maximum radiation is at an elevation of 90° and there is very little radiation at low angles. This can be an advantage for short distance communication such as daytime working on the lower frequency bands. For long distance communication, it is desirable that the maximum radiation from the aerial system (this includes ground and other nearby conductive objects) should be at a low angle relative to the horizon. This is usually achieved by using one or more ground mounted verticals or by placing a horizontal aerial as high as possible above ground. The optimum height for horizontal arrays ranges from 0.5 wavelengths to several wavelengths above ground. This is the reason why verticals are so popular with DXers on the lower frequency bands. Very few amateurs are able to place an aerial more than 20-30 meters above ground.

DIRECTIONAL AERIALS. In the late 1800s, experiments by Heinrich Hertz and other pioneers in the field of electro-magnetics showed that radio waves could be reflected by conductive objects and focused in one direction by placing the transmitting dipole adjacent to a concave reflector. These important discoveries would eventually lead to the development of highly directional aerials used for radar and radio astronomy. There are many different types of directional aerial. Some are based on arrays of directly driven elements like the HB9CV beam [**55**] or phased arrays of verticals as used in the 4-square array. Other designs use parasitically driven elements as used in the Yagi-Uda array or the multi-element quad beam. Some UHF and microwave aerials use a large parabolic dish or a pair of flat conductive surfaces arranged in a V shape as a passive reflector. It would not be possible to describe all of the many different types of beam aerial in a few pages, but we will take a brief look at some of the most common types.

The Yagi-Uda Array was developed by Shintaro Uda and Hidetsugu Yagi in 1926. Yagi-Uda is usually abbreviated to Yagi. The Yagi array usually consists of a single driven element and one or more parasitic elements. A simple half-wave dipole or folded half-wave dipole is often used as the driven element. Parasitic elements at the rear of the array (relative to the direction of maximum radiation) are known as reflectors. Parasitic element at the front of the array are known as directors. The simplest form of Yagi consists of a single driven element and one parasitic element. The parasitic element in a two element Yagi can be either a director or a reflector. Most two elements designs use a reflector. The phase and magnitude of the RF current in a parasitic element depends on the element length and the spacing between it and other elements in the array. In a typical two element design, the reflector is a few percent longer than the resonant driven element and the element spacing is 0.15 to 0.25 wavelengths. Even this most basic Yagi can have a maximum gain of more than 6dBi (decibels relative to an isotropic radiator). **Figure 5.5** shows a NEC2 plot of the radiation pattern from a simple two element VHF Yagi. The driven element length is 930mm and the reflector length is 1020mm. Element spacing is 300mm or approximately 0.15 wavelengths. At 144.0MHz, the maximum gain is 6.58dBi and the F/B (front to back) ratio is about 10dB. It is possible to optimise the element lengths and spacing for even greater maximum gain or F/B ratio. The old adage 'you don't get owt for nowt' can be applied to Yagi aerials. The designer will often have to choose between several conflicting requirements. Optimising a Yagi design for the maximum possible gain will result in a reduction in F/B ratio and/or bandwidth. Changes to the element lengths and spacing will also affect the feedpoint impedance of the driven element. Yagi designs that are highly optimised for maximum gain tend to have a very low feedpoint

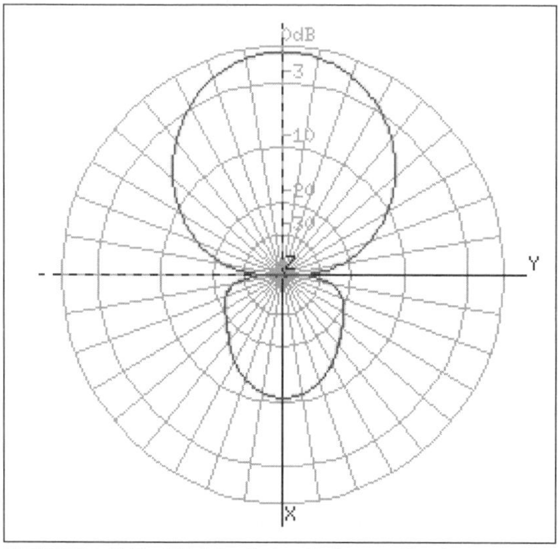

FIGURE 5.5: NEC2 plot of the radiation pattern from a simple two element VHF Yagi.

impedance which will result in increased resistive losses in the aerial elements. Very low, very high or extremely reactive feedpoint impedance will also lead to increased losses in the feed system.

Figure 5.6 shows the layout of a typical three element Yagi. The driven element is a simple centre-fed resonant half-wave dipole. The reflector is about 4-5% longer than the driven element and the director is about 4-5% shorter than the driven element. The spacing between the driven element and the reflector is typically about 0.2 wavelengths. The spacing between the driven element and the director is often chosen to obtain a desired value of feedpoint impedance rather than the maximum possible gain or F/B ratio.

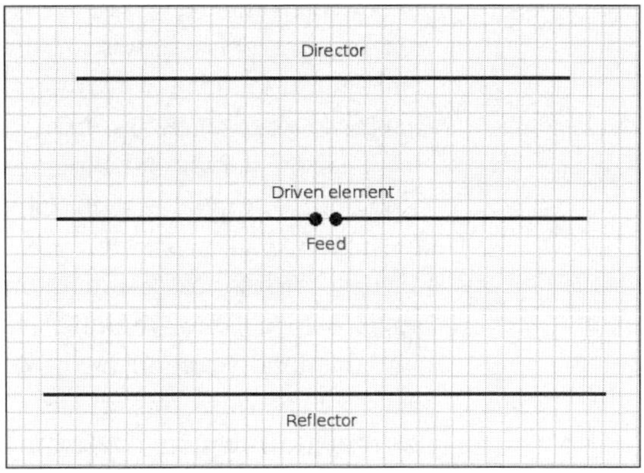

FIGURE 5.6: Layout of a typical three element Yagi.

While there is no upper limit to the number of elements that can be used in a Yagi, practical considerations of mechanical strength, high costs and the difficulties involved in rotating a very large array mean that HF Yagis are usually limited to about 5-7 elements. At VHF 15-18 elements become practical and at UHF and above it is possible to make Yagi aerials with dozens of elements. **Photo 5.5** shows a home made 30-element Yagi for the 23cm band. The gain of this array is around 19dBi. Commercial beams with more than 50 elements are available for 23cm and 13cm. On the higher microwave bands, parabolic dishes offer greater gain than a typical Yagi array. Multiple Yagis can be 'stacked' vertically or 'bayed' horizontally to create larger arrays. Four Yagis stacked and bayed into a square formation is a popular configuration for amateur VHF and UHF EME (moonbounce) arrays.

PHOTO 5.5: Home made 30-element Yagi for 23cm.

OTHER TYPES. The Yagi is by far the most popular type of directional beam, but there are several other types in common use. Quad beams consisting of a full wavelength resonant loop as the driven element and one or more parasitic elements are very popular on the higher HF bands. A two element 'Cubical Quad' beam has two square loops mounted on a relatively short boom. The boomless 'Spider Quad' type of construction is particularly suitable for multi-band beams. The EI7BA Multiband Cubical Quad [**56**] is a prime example of this type of beam. There is no requirement for the loops in a multi-element quad to be square. Triangular loops can be used to make a 'Delta Loop' beam and circular loops are often used in UHF and SHF 'Loop Yagi' beams.

V-BEAMS AND RHOMBICS. Another form of directional beam can be made from very long wire aerials. A pair of long wires can be arranged in a V configuration to make a bidirectional V-beam. A pair of V-beams can be connected back-to-back to form a Rhombic aerial which is named after its rhombus shape. Like the V-beam, an unterminated rhombic is bidirectional. If one side of the rhombic is terminated by a resistor it becomes unidirectional. The rhombic works well over a wide frequency range. It is quite easy to make and because of it's broadband nature, it's dimensions are not very critical. High gain, low frequency rhombics are much too large to fit in the average garden. The diamond shape of the rhombic means that they need to be supported at four points which makes them almost impossible to rotate. It is possible to reverse the direction of a terminated rhombic by swapping the position of the feedpoint and termination resistor.

There are several space saving variants of the Yagi. Most of them are two or three element Yagi beams with shortened elements. The shortening of the elements usually involves some form of bending of the wire elements to reduce the width of the array. One of the most popular compact two element beams is known as the Moxon Rectangle. This is a two element Yagi which has the ends of the wire elements bent inwards so that they almost meet. The Moxon Rectangle is based on the VK2ABQ wire beam. **Figure 5.7** shows the layout of the VK2ABQ. The VK2ABQ is square so that the width of the aerial is exactly the same as the total length of the sides. Moxon **[57]** increased the width of the array and shortened the boom length accordingly which results in a rectangular shape which has a slightly larger turning circle than the VK2ABQ. Moxon's design was been analysed and further refined by L B Cebik W4RNL [**58**]. Windows PC users can use MoxGen software by Dan Maguire, AC6LA, to design a Moxon Rectangle [**59**]. **Photo 5.6** shows a 10m Moxon used at EI9GQ during the peak of the last sunspot cycle.

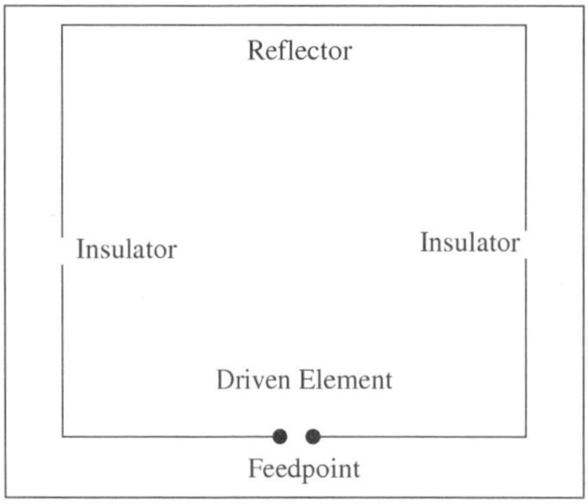

FIGURE 5.7: Layout of the VK2ABQ wire beam.

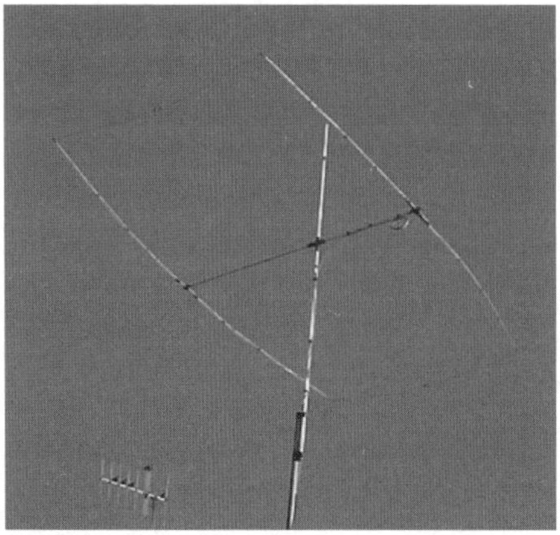

PHOTO 5.6: 10m Moxon antenna used at EI9GQ.

BEAM CONSTRUCTION. Any good DIY store should be able to supply most of the parts for a yagi or quad beam. I prefer to use non-conductive materials for the boom of VHF and UHF beams. This greatly simplifies the design because there is no need to apply a boom correction factor to the element lengths. It is also much easier to model a beam with a non-conductive boom using NEC or similar modelling software. It is possible to use a conductive metal boom without applying any correction to the element lengths provided that the boom diameter is small compared to the length of the elements and that the elements are offset from the boom by mounting them on plastic stand-off insulators. Some types of electric fence insulators can be used for this purpose. The boom to element spacing should be at least equal to or greater than the element radius (0.5 x element diameter). **Photo 5.7** shows a 2.4GHz Yagi on a non conductive plastic conduit boom.

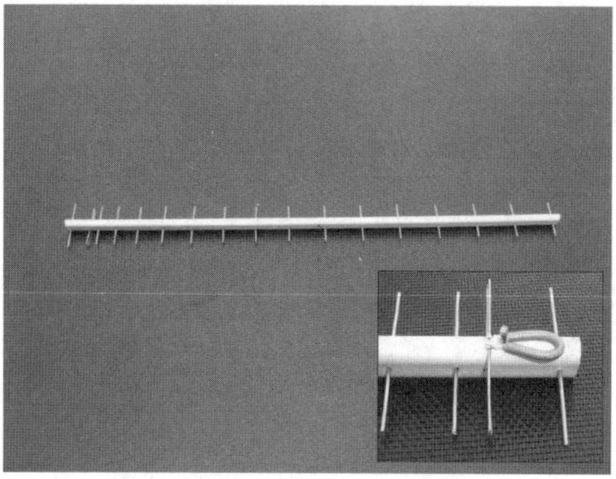

PHOTO 5.7: 2.4GHz Yagi on a non conductive plastic conduit boom.

PVC electrical conduit can be used as spreaders for VHF/UHF quad and delta loop beams. Standard types of conduit are not strong and rigid enough for larger HF quads. Fibreglass spreaders are ideal for HF quad beams, but they are quite expensive to buy and not readily available in this part of the world. Bamboo is a suitable low cost alternative. The Moxon rectangle in Photo 5.6 uses bamboo rods from the local garden centre as the element supports **[60]**.

13-ELEMENT 70CM YAGI. This project is a 13-element UHF Yagi. This beam is based on the classic DL6WU design. All of the elements (see **Table 5.1** for dimensions) are made from 6mm aluminium rod. I used a non-conductive timber boom. The maximum gain of this beam is about 15dBi, F/B ratio is just under 20dB and the horizontal 3dB beamwidth is approximately 35°.

6mm solid aluminium rod was used for all of the elements. Aluminium tubing with the same outside diameter can be used instead. I used the solid rod because it was considerably cheaper than hollow tube! The boom is a 2.4 metre length of 46x19mm timber from a local DIY store. It took quite a while to find a perfectly straight length and even longer to explain to the store assistant why it had to be perfectly straight. The reflector and last director are mounted 20mm from the ends of the boom. The boom was cut to a total length of 2280. It is best to do this after you have drilled the element mounting holes. It is very easy to cut off the surplus timber at the end of the boom, but not so easy to put it back if you make a mistake. The element positions were marked on the boom using a felt tip pen. After double checking the accuracy of the element positions, the element holes were drilled out using a 6mm drill. Ideally, a drill stand should be used to endure

Element	Spacing	length	Element	Spacing	length
Ref	0.000	337.0	D6	1040.9	279.8
D.E.	138.8	330.5	D7	1259.5	277.3
D1	190.8	397.8	D8	1488.5	275.2
D2	315.7	293.6	D9	1727.9	273.3
D3	464.9	289.6	D10	1977.8	271.5
D4	638.4	285.9	D11	2238.0	270.0
D5	832.7	282.6	All elements 6mm diameter.		

Table 5.1: DL6WU Yagi element length and spacing.

that the element holes run perfectly straight through the boom. It is possible to use a hand-held drill if you take great care to keep the drill vertical while you are drilling the holes. I used a bench vice to hold the elements while they were cut to the correct length using a hacksaw. The element length must be correct to within ±2mm. I found that I could easily trim the elements to the exact length required by cutting them about 0.5mm too long and filing the ends until the element was exactly the right length. Sub-1mm accuracy is easily achieved using a good engineers' rule.

PHOTO 5.8: How the Yagi elements are mounted.

To avoid any possibility of getting them mixed up, the elements should be fitted to the boom as soon as they are cut. The elements are such a tight fit in the boom that I had to use a block of timber as a mallet to tap them into place. **Photo 5.8** shows how the elements are mounted.

The driven element is a folded dipole made from 6mm aluminium. The element was bent into shape by bending the aluminium around a steel bar. **Figure 5.8** shows the details of the driven element and 4:1 coax balun. The ends of the aluminium at the dipole feedpoint were partially flattened using a heavy hammer and drilled using a 4mm drill. A pair of M3 nuts and bolts were used to attach solder tags to the feedpoint. The 4:1 balun is made from a half-wave length of 50Ω coax cable. I used a length of RG58 with a solid dielectric.

PHOTO 5.9: The finished DL6WU Yagi.

The length required is ((300/f)*VF)/2
((300/432.2)*0.66)/2 = 229mm

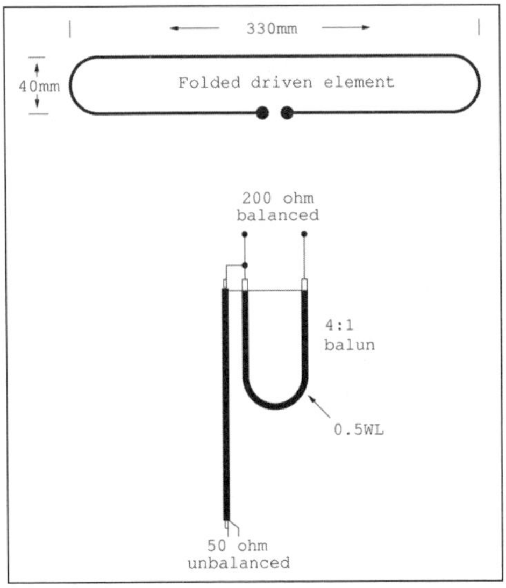

FIGURE 5.8: The driven element and 4:1 coax balun.

I used an N-type socket mounted on a steel L bracket to attach the 50Ω feeder to the balun and driven element. As I am still testing this design, I have not made any real attempt to waterproof the coax balun.

The Yagi was mounted on a fibreglass pole at a height of 11m above ground. The feeder is a 12m length of Westflex 103 50 ohm coax. **Photo 5.9** shows the finished Yagi. The receiver is a simple home made 70cm to 10m converter and a HF transceiver as the 10m IF. The first tests were performed in February when 70cm activity was

even lower than usual and tropospheric conditions were less than ideal. After about one hour of listening on 432.934MHz, the Bristol 70cm beacon GB3BSL appeared out of the noise! The beacon in IO81QJ is 400km from my location. A recording of the test can be found at:
http://homepage.eircom.net/~ei9gq/Bristol.mp3

I was unable to test the SWR seen at the feedpoint because I don't have a 70cm transmitter. **Figure 5.9** shows a NEC2 plot of the free-space horizontal radiation pattern of the 70cm Yagi **[61]**.

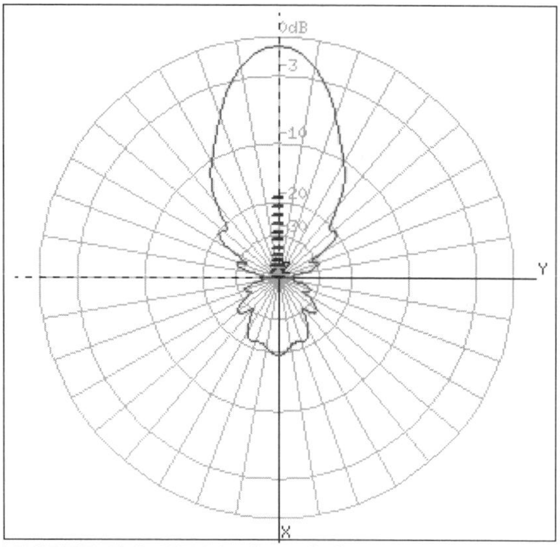

FIGURE 5.9: NEC2 plot of the Yagi's free-space horizontal radiation pattern.

ATUs and Aerial Feed Systems

WHAT IS AN ATU AND WHY DO I NEED ONE? An ATU is an impedance matching device which is used between the RF input/output connector of a transmitter/receiver and an aerial system. Most transmitters and receivers are designed to use a 50Ω unbalanced co-axial connection for the aerial input/output. The ATU allows us to match an aerial system of arbitrary impedance to 50Ω, 75Ω or any other value that we require. Many of the LC impedance matching networks that we have used in previous homebrew projects are quite suitable for use in an ATU. PI, T and L networks are all widely used in ATUs. In fact, it would only be a slight exaggeration to say that any random collection of inductors and capacitors could be connected together in any random configuration to make some form of ATU.

As with most things that are associated with aerial systems, there is some controversy over the name ATU. Aerial /Antenna Tuning Unit or AMU (aerial/antenna matching unit) is often considered to be a misnomer because in most cases, the tuner is not used to tune the aerial to resonance. It is simply an impedance transforming device which matches whatever impedance happens to appear at the shack end of the feeder to 50Ω or some other preferred value. Some amateurs prefer to use a more technically accurate term like ASTU (aerial system tuning unit) or Transmatch.

There are many situations where an ATU is not required. Some aerials have a feedpoint impedance which is reasonably close to 50Ω. A quarter-wave ground plane or a centre-fed half-wave dipole are typical examples. Many aerials which have a feedpoint impedance that is radically different from 50Ω include a built-in matching system so that the impedance seen at the feedline is close to 50Ω. This matching system can be made from sections of transmission line as used with the G5RV doublet and the J-Vertical or from lumped

LC components. Some matching systems like the gamma-match use a combination of transmission line and lumped LC components.

Most of the aerial systems used by radio amateurs are inherently narrow bandwidth devices. The centre-fed half-wave dipole as discussed in last month's Homebrew has a free-space feedpoint impedance of about 73Ω at resonance. In the real world, the proximity of ground and other conductors will have some effect on the feedpoint impedance and in many cases the actual feedpoint impedance will be even closer to 50Ω. At its resonant frequency, the inductive and capacitive reactances of our dipole effectively cancel each other out so that the impedance at the feedpoint appears to be a 'real' resistance. This is only true at the resonant frequency. If we increase the operating frequency the electrical length of the dipole will be greater than 0.5 wavelengths and the feedpoint will show an inductive reactance. If we decrease the frequency, the electrical length of the dipole will decrease and it will present a capacitive reactance to the feed system. As a general rule, an electrically short aerial will present a capacitive reactance and an electrically long aerial will present an inductive reactance. The resistance seen at the feedpoint will vary with the electrical length of the dipole. A dipole which is very short compared to a half wavelength will have a very low feedpoint resistance and a very large capacitive reactance. As the length of a dipole is increased beyond a half wavelength, the feedpoint resistance will increase until we reach a point where the total length of the aerial reaches one wavelength. At this point, our dipole has become a full-wave doublet which is effectively a pair of resonant half-wave dipoles fed in phase. The feedpoint resistance is several thousand ohms and because the doublet is resonant, the feedpoint reactance is zero. **Figure 5.10** shows a NEC2 plot of the gain, front to back (FB) ratio, SWR and impedance seen at the feedpoint of a centre fed doublet. The total length of the doublet is 39 metres and the wire diameter is 2mm. This is similar to the HF doublet described previously. The NEC2 software has been configured to calculate the gain and impedance at 100 different frequencies starting at 3.0MHz and ending at 15.5MHz. Graph A shows gain as a dotted line and FB ratio as a solid line. Since the doublet is a bi-directional aerial, the FB ratio is al-

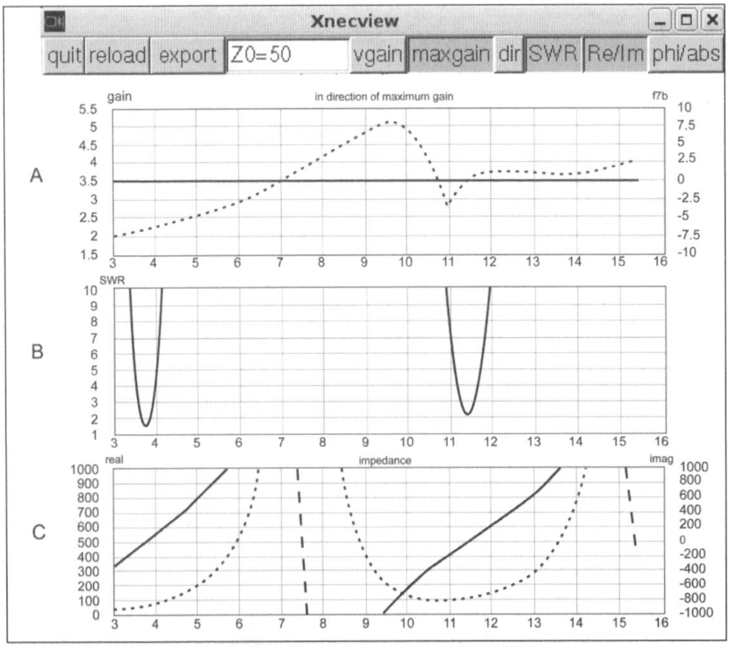

FIGURE 5.10: (A) NEC2 plot of gain (dotted line), front to back ratio (solid line), (B) SWR (solid line), (C) resistive (real) impedance (dotted line) and reactive impedance (solid line).

ways 0dB. Graph B shows the SWR (standing wave ratio) that would appear on a 50Ω feeder connected to the feedpoint. Graph C shows the real (resistive) part of the feedpoint impedance as a dotted line and the imaginary (reactive) part of the feedpoint impedance as a solid line.

If you look at the left side of the impedance graph, you will see that the reactance plot crosses the zero line at about 3.75MHz. This is the half-wave resonant frequency of the doublet and coincides with the lowest point on the SWR plot. Note that a negative value of reactance signifies a capacitive reactance and a positive value signifies an inductive reactance. You will also notice that the gain is approximately 2.15dB at this frequency. All of these figures are in the expected range and it seems that our NEC2 model is reasonably accurate. Apart from the half-wave resonant frequency, there are several other points of interest on the graphs. The full-wave resonance can be seen at around 7.5MHz where the aerial operates as a co-linear pair of half-wave dipoles. The maximum gain broadside to the doublet is 3.5dBi (1.36dBD) at 7.1MHz. The feedpoint resistance at 7.1MHz is very high and because the reactance changes very rapidly with frequency near the full-wave resonance, there is considerable reactance at 7.1MHz. If we were to use a 50 or 75Ω co-ax feeder, the SWR on 40m would be about 90:1. This would result in very inefficient operation because of the very high feeder losses. The next point of interest is the peak in maximum gain at about 9.5MHz. At this frequency, each side of the doublet is 5/8 (0.625) wavelengths. This is the length that provides maximum gain broadside to the doublet. If we increase the frequency much beyond this point, the two main lobes of the radiation pattern will start to break up into several smaller lobes. A 2x 5/8 wave doublet is often referred to an EDZ (extended double Zepp) and has a maximum gain of just over 5dBi. The next point of interest is around 11.25MHz where the doublet is 1.5 wavelengths (or 2x ¾ wavelengths). Like the half-wave resonance the feedpoint is at a low voltage, high current (voltage node) point and the impedance is a reasonably good match to 50 or 75Ω. Note the sharp dip in the SWR at this point. At four times the half-wave resonant frequency, the doublet is four wavelengths long and the feedpoint impedance is very high. This point can be seen at just over 15MHz on the impedance graph. The doublet can be considered as four half wave dipoles split into two pairs. At 15MHz, the current in each pair of dipoles is not in phase so the doublet will not operate as a co-linear array. The radiation pattern splits up into four lobes and resembles a four leaf clover - see **Figure 5.11**.

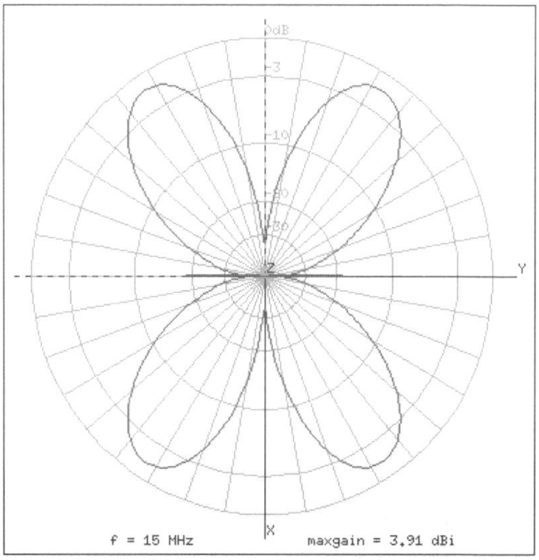

FIGURE 5.11: The clover-leaf radiation pattern of the doublet at 15MHz.

The feedpoint impedance will be low at the half-wave resonant frequency and at odd multiples of this frequency. The feedpoint impedance will be high at even multiples of the half-wave resonance. One significant point to note is that the maximum gain at every frequency above 3.75MHz is greater than the 2.15dBi gain of a half-wave dipole. This tends to support the general principle that bigger is better (at least in some directions) where aerials are concerned.

The gain graph in Figure 5.10 shows clearly that this doublet would make an excellent multi-band aerial. It would work very efficiently on 80m as a simple dipole. It offers gain over a dipole on 40m and considerable gain on 30m where it is just over 2x 5/8 wavelengths. On 20, 17, 15, 12 and 10m the radiation pattern breaks up into multiple lobes, offering useful gain in several directions. To get this HF doublet working well on all HF bands, we will need to find an efficient way of feeding it.

A simple co-ax feed will only work well on 80m and exact odd multiples of its resonant frequency. Since it is resonant on 3.75MHz, none of these resonances fall within any of the HF amateur bands. The situation is even worse on even multiples of 3.75MHz where the feedpoint impedance is very high. There are several approaches to solving this problem. We can place parallel-resonant traps at strategic points along the wire which will isolate the outer parts of the doublet at the resonant frequency of the traps. This will mean that the doublet will operate as a half-wave dipole on each band. This approach is also used for multi-band trapped verticals. The most common type of trap consists of a single coil connected in parallel with a high voltage fixed capacitor. The other type of trap in common use is made from a coiled quarter-wave length of co-axial cable. I try to avoid using traps in my aerials because I feel that they usually add unnecessary complexity, they result in a system which is difficult to tune and they are prone to failure caused by excessive RF voltage and rainwater. It is possible to use sections of transmission line as open or short circuit stubs to match a 50Ω feeder to a high impedance aerial feedpoint. The J-Vertical is a typical example of such an arrangement. The J-Vertical is really just another name for a Zepp (short for Zeppelin) aerial. Supposedly, the quarter-wave matching section of the Zepp was used to keep the high voltage ends of the trailing half-wave wire far away from the explosive hydrogen gas used in early Zeppelin airships.

To achieve the lowest possible losses in the feed system, the feedline should be terminated by a resistance which is equal to its characteristic impedance. Under

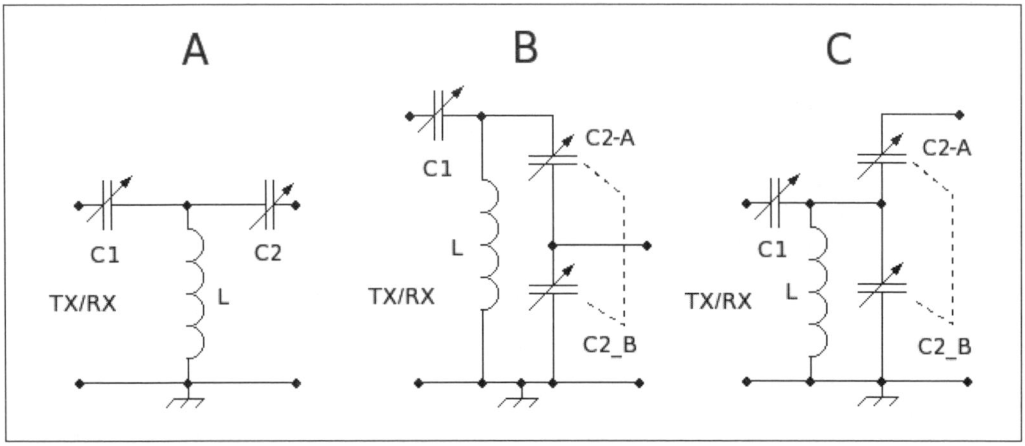

FIGURE 5.12: Different ATU configurations (see text).

these conditions, the SWR on the line will be 1:1 and apart from the small percentage wasted as line losses and resistive losses in the aerial wire, all of the transmitter power will be radiated as radio waves. Unfortunately, our doublet can only satisfy these conditions on 80m. One of the best ways of feeding a multi-band doublet is to use low-loss balanced line between the transmitter/receiver and the aerial feedpoint. Wide spaced open wire line has a characteristic impedance of about 600Ω and has extremely low losses of only a few dB per kilometre when it is properly terminated. The length of open wire feedline used in a typical amateur station would have negligible losses of a small fraction of 1dB when correctly terminated. Even when such a line is not properly terminated, the line losses will usually be acceptably low. The impedance seen at the shack end of the line will vary widely from band to band. We will have to use some kind of matching network to match our 50Ω transmitter/receiver to this unknown line impedance. Because the doublet and feedline are balanced with respect to ground, we will need to use a matching system that provides a balanced connection to the feedline. The unbalanced to balanced transformation can be performed by the ATU or alternatively, we can use an unbalanced ATU in combination with a balun (balanced-to-unbalanced transformer).

PRACTICAL ATU CIRCUITS. The simplest practical ATU is a simple L-network consisting of a single capacitor and a single inductor. A low-pass type of L-network with a series L and shunt C or the high-pass equivalent with a series C and shunt L are both suitable for this job. This type of matching network is often used in automatic ATUs where banks of inductors and capacitors are switched by relays. The number of possible values of C or L is 2^N, where N is the number of relays used. One problem with the L-network is that it can only be used to match a low impedance to a high impedance or vice versa. To accommodate impedances that may be greater or less than 50 ohms, it is necessary to use another relay or switch to swap the network low-Z and high-Z connections. The addition of a third reactive element allows us to build a much more flexible matching network. The addition of a second capacitor allows us to make a low-pass PI network or a high-pass (CLC) T-network as shown in **Figure 5.12a**. This type of T-network is very widely used in both home made and commercially made ATUs. If you have a dual-gang capacitor, you could build the 'Ultimate Transmatch' as shown in **Figure 5.12b** or the 'SPC Transmatch' as shown in **Figure 5.12c**. Both of these matching networks are derived from the T-network and offer greater selectivity than the simple T-network.

BALANCED TUNERS. Some types of ATU provide a balanced connection to the aerial. **Figure 5.13** shows a link-coupled tuner which is designed to work with balanced feedline. C2-A and C2-B are a dual-gang tuning capacitor. The capacitor body and rotor plates of C1 and C2 are grounded.

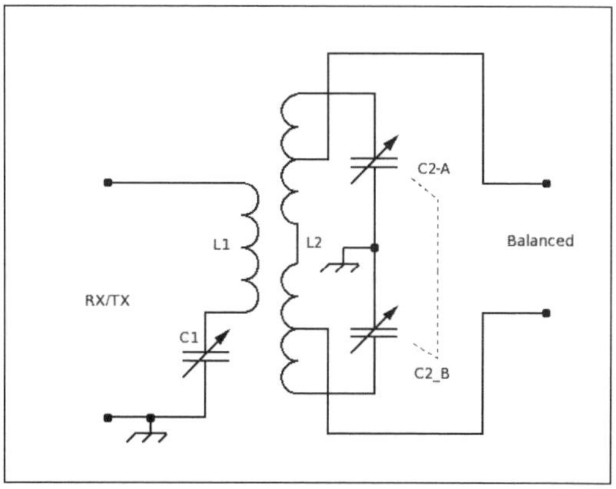

FIGURE 5.13: Link-coupled tuner intended for use with a balanced feedline.

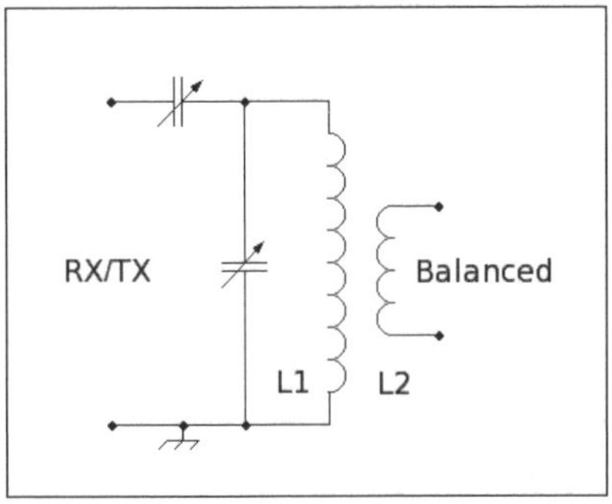

FIGURE 5.14: A simpler form of balanced tuner.

This is very convenient when the tuner is built in a metal enclosure. There is no need to isolate the capacitors from ground as required by some other designs. Band switching of this type of tuner can be quite complicated because of the need to adjust the value of both L1 and L2 and to find the optimum tapping point for the balanced feeder on L2. This circuit was very popular in the days when open construction and crocodile clip connections were the norm.

Figure 5.14 shows a simpler form of balanced tuner. This circuit is also link-coupled and it is easily adapted to unbalanced operation by simply grounding one end of the coupling link. Band switching is achieved by using a two-pole switch to connect to taps on L1 and L2.

A T-MATCH ATU. This next project is a simple T-Match ATU. A balun is included to allow operation with balanced feedlines. I have been using this tuner for the last 17 years so it is safe to say that the design has been well tested. I have upgraded some of the components over the years. The original PVC insulated wire coil which was wound on a plastic former has been replaced by a very heavy-duty inductor. The balun was also upgraded a few years ago. The basic design remains the same as the original **[62]**.

When there is a severe mismatch between the 50Ω input of the tuner and the aerial output of the tuner, it will be necessary to adjust the ATU so that it is operating at a high loaded Q. This will tend to lead to large losses in the tuner. The only way to avoid these losses is to use very low loss components. Large transmitting type air-spaced capacitors are high-Q, low loss devices. The inductor(s) used in the matching network are a key part of an ATU. To keep resistive losses to a minimum, the coils should be made from thick copper wire. A wire diameter of several millimetres is recommended. The band switch(es) should be capable of withstanding very high RF voltages. Large ceramic switches with a current carrying capacity of many amps and the ability to withstand peak voltages of several thousand volts between the switch contacts and ground are ideal for this job.

CONSTRUCTION. I built the ATU in a partially open enclosure (if that's not too much of a contradiction) made from copper-clad PCB laminate. High power transmitter components like air-spaced capacitors and ceramic switches are not readily available from normal component suppliers. Suitable second-hand capacitors, coils and switches can be found from time to time at radio rallies and on-line auction sites like eBay. Coils are easy enough to make from large diameter copper wire. Brand new air spaced capacitors are usually prohibitively expensive, but you can buy air-spaced capacitors in kit form. Keep an eye open for bargains at rallies, car boot sales, scrapyards and anywhere you might find surplus radio or medical equipment. A well stocked junk-box is essential for the QRO equipment builder.

The main coil in my ATU was pulled from an old MF transmitter. It spent several years sitting on a shelf until I found a use for it. The coil is made from 5mm wide, 1.5mm thick, silver plated copper. For all practical purposes, this coil can be considered lossless and utterly indestructible. **Photo 5.10** shows a close-up view of the coil. This inductor is probably overkill for amateur power levels. A simple coil made from 2.5mm earth wire wound on a 50mm plastic pipe or other suitable former can be substituted for L1. If you want to cover the 160m band, about 40 turns should be enough. This will give an inductance of around 30µH. 30 turns should be adequate to cover 80m-15m. The band switch was scrounged from the junk-box of a local amateur (thanks John!). Band switching is very easily arranged because only a single-pole switch is required. I found that my large MF coil had too much stray capacitance for use on the 10m band. This problem was solved by using a separate coil for 10m and 12m. The 10m coil is made from 3.5 turns of 2mm silver plated wire with a diameter of approximately 40mm. This coil is self supporting so no coil-former is required. The two air-spaced tuning capacitors were recovered from a scrap marine transmitter. One has a maximum capacitance of 500pF the other has a maximum capacitance of 300pF. An extra complication of the CLC T-match is that both sides of both capacitors must be isolated from ground. This is easily arranged because the holes for the mounting screws are in the ceramic front plate of the capacitors. This means that the capacitors can be bolted to the front panel without connecting them to ground. If you are using metal-bodied capacitors, you will have to ensure that they are isolated from ground. Many constructors use a Perspex mounting plate for this purpose. The 50Ω unbalanced connection to the rig is via a standard SO239 type co-axial socket. The aerial connection for both balanced and unbalanced feeders is via screw-on connections on the back panel of the ATU. See **Photo 5.11** for details.

PHOTO 5.10: The ATU main coil, made from silver plated copper.

BALUN. I use a 4:1 balun between the output of the T-match and the balanced open wire line terminals on the back of the ATU. This is considered by many amateurs to be the 'wrong way' to do it, but I have found that it works very successfully in my situation. This same method is used in most commercially made ATUs. The balun consists of 6 to 10 turns of bifilar wound PVC insulated wire wound on a side-by-side pair of ferrite RFI suppression cores. I don't have the specifications of the large ferrite cores used in my original balun. Some of the larger RFI suppression cores like the N90AB from Maplin look like a suitable candidate for the job. To test the performance of the Maplin cores in a broadband balun, I made a 4:1 balun using 6 turns bifilar wound on a pair of N91AB cores that were left over from the HF PA project. This balun works well at all frequencies from 1.8-30MHz. It gets

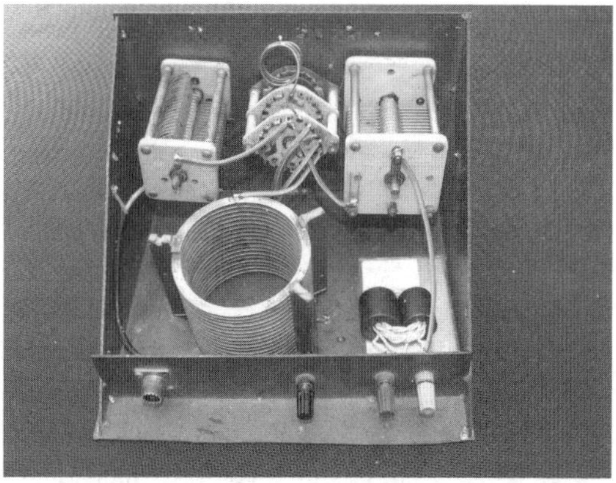

PHOTO 5.11: The completed CLC T-Match.

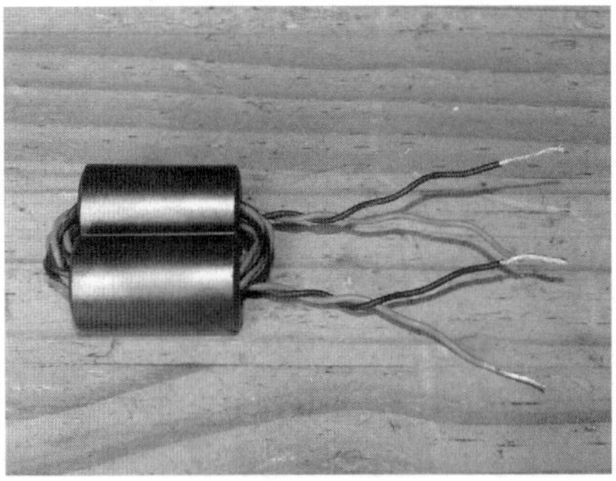

PHOTO 5.12: Testing the Maplin cores in a balun configuration.

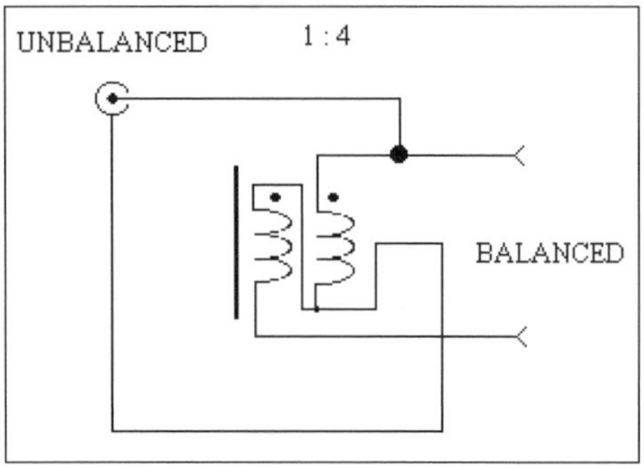

FIGURE 5.15: The winding arrangement of the balun.

lukewarm at 100W PEP levels and can survive 200W+ at 100% duty cycle for two minutes without getting too hot. This balun is shown in **Photo 5.12**. **Figure 5.15** shows the arrangement of the balun windings. The original balun with the large cores has been tested at up to 400W.

ATU Project and a Simple SWR Meter

THE CIRCUIT. **Figure 5.16** shows the T-Match ATU circuit. A separate coil (L2) is used for the 10m and 12m bands. L1 is used for the lower frequency bands. A single-pole switch is used for band switching. The optional 4:1 balun is shown within the dashed lines.

The amount of inductance required for each band will depend to some extent on the impedance that is seen at the shack end of the feedline. Unless you are fortunate enough to have a roller-coaster or similar type of variable inductor, you will need to find the optimum tapping point on L1 for each band. As a general rule of thumb, the ATU coil should have enough inductance to provide an inductive reactance which is several times greater than the normal working impedance of the ATU. For a typical 50 ohm system, the coil should have a reactance of several hundred ohms at the lowest working frequency. The formula $L = X_L/(2pi*f)$ can be used to calculate the inductance required for a given value of X_L (where L is in µH, X_L is in ohms and f is in MHz). For example, to cover the 160m band, 50/(6.283*1.8) gives a value of 4.42µH for 50ohms at 1.8MHz. A coil with an inductance of 20-30µH will be adequate for use in a 160m T-Match. This is a typical value for commercially made air-cored coils and roller-coaster inductors. 30-40 turns of 2.5mm insulated wire on a 40-50mm diameter former makes a good home made ATU coil. **Photo 5.13** shows a home made coil. The coil former is a length of plastic pipe and the wire is 2.5mm earth wire.

TESTING. Aligning of the ATU is a relatively easy job. It will be even easier if you have access to specialised test instruments like an SWR analyser, a RF bridge or a noise bridge and HF receiver. Since these are not standard items in the average amateur shack, I will assume that you have only a transceiver, an SWR meter and a dummy load and/or aerial.

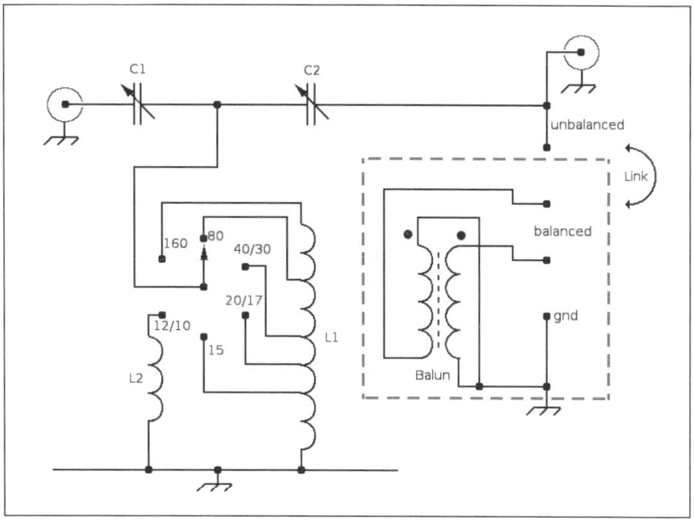

FIGURE 5.16: The T-Match ATU circuit.

Connect the transceiver to the 50Ω input of the ATU. Unless the transceiver has a built-in SWR meter, you should place your SWR meter in line between the transceiver and the ATU. Connect the aerial to the output terminals of the ATU. If a balanced feedline is used, you should use the 4:1 balun transformer. Connect one end of L1 to the metal chassis of the ATU. As my ATU enclosure is made from copper clad PCB laminate, this was just a simple matter of soldering the wire directly onto the case. If you are using a metal enclosure you can use a small screw, nut and flat washers to fix a solder tag onto the bottom of the case. Make sure there is a good low-resistance connection to ground. The other end of the coil should be connected to the 160m position on the band switch. Place the two capacitors at the middle of their adjustment range, select the 160m position on the band switch and tune the receiver to 1.9MHz. You should find that you can peak the band noise by alternately adjusting C1 and C2. Tune for maximum received noise, note the positions of the capacitors and then connect a dummy load to the output of the ATU.

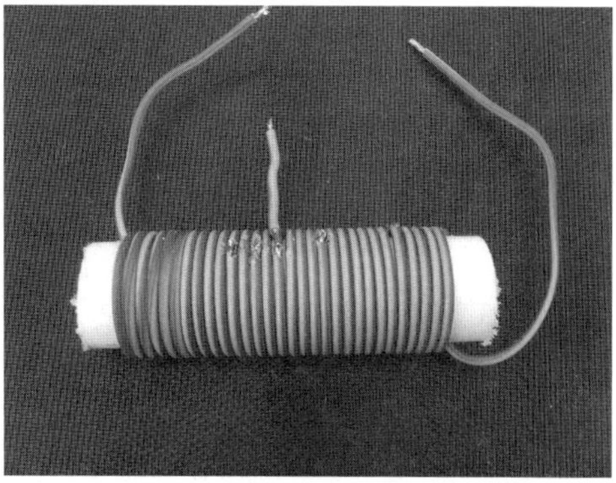
PHOTO 5.13: Home made ATU coil made from 2.5mm earth wire wound on a length of plastic pipe.

Most, if not all, modern transceivers have a protection circuit which reduces the output power if the transmitter is operated into a mismatched load. These SWR protection circuits are generally quite reliable, but they are not always completely foolproof. You should operate the transmitter at reduced power while you are adjusting the ATU. Operating a 100W transmitter at an output of 10W or less will greatly reduce the risk of damage. Keep your transmission periods short while you are adjusting the ATU.

Set the output of the transmitter to a suitably low level and make a brief test transmission. With a little bit of luck, you may find that the capacitor positions from the receive noise test will result in a reasonably low SWR. Alternately adjust C1 and C2 until a perfect 1:1 SWR is achieved. If you find that a good match is not possible, you will need to adjust the number of turns on L1. This is very easy to do when bare copper wire is used for L1, but it can be a bit tedious if the coil is wound from insulated wire. You will need to cut the insulation from the wire at regular intervals along the coil and tin the bare copper with solder so that the wires from the band switch can be soldered to the coil. You can see that the coil in

Photo 1 bears the scars from several encounters with the soldering iron. Once you have found the optimum amount of inductance for 160m, it is usually quite easy to find the best tapping points for the higher bands. The large coil in my T-Match is 17 turns with a diameter of 95mm and coil length of 90mm. The inductance of this coil is just under 20μH which is just about adequate for 160m. It works fine above 1.9MHz, but both tuning capacitors need to be almost fully meshed to achieve a good match at 1.82MHz. The full 17 turns are used for 160m, 9 turns for 80m, 6 turns for 40m/30m, 4 turns for 20m/17m and 3 turns for 15m. The 10m/12m bands are covered by a separate coil which is made from 3.5 turns of 2mm wire with an inside diameter of 40mm. The 10m/12m coil is self supporting and mounted directly on the band switch. I did a few tests with a home made coil which was wound from insulated wire on a plastic pipe. The pipe outside diameter is 43mm and the wire is standard 2.5mm green/yellow earth wire. 32 turns gave good results across the entire 160m band. 14 turns worked well on 80m, 10 turns on 40m, 6 turns on 20m/17m and 5 turns on 15m. The tuner was able to provide a good match to the open-wire feeder of my HF doublet on all nine HF bands.

Like other controlled-Q matching networks, the T-Match can provide a good match at more than one value of coil inductance. Using the smallest value of inductance that will provide a good match at the frequency in use will result in the lowest possible losses in the tuner.

TRANSMISSION LINES AND SWR. When a transmission line is terminated by a pure resistance which is exactly equal to its characteristic impedance, all of the power available at the load end of the line will be dissipated by the load. There will be no reflected power, and no standing waves will appear on the line. A properly terminated line is sometimes referred to as a flat line, a matched line or a non-resonant line. It is sometimes useful to visualise such a line as a lossless and infinitely long transmission line. Energy fed into this infinitely long line will continue to propagate along the line forever. Power fed into the line will simply disappear and never be seen again. The impedance seen at the generator end of such a line would always be equal to the characteristic impedance of the line.

The situation described is not possible in the real world. A practical line will have a finite length and it will often be terminated by a load which is not a perfect match for the line impedance. All transmission lines will show some loss due to resistive losses in the conductors and dielectric losses in the insulator between the conductors. Voltage or current imbalance on the line will lead to line losses by electromagnetic radiation from the line. This can regarded as a problem in most situations or as a useful feature in a few special cases.

Impedance is often expressed in the form R + jX, where R is the resistive (real) part of the impedance and j is the reactive (imaginary) part of the impedance. Positive values of j indicate inductive reactance, negative values of j indicate a capacitive reactance. In the case of a 50Ω line, a perfect match can only be achieved if the terminating impedance is a pure resistance of 50Ω (50+j0). If the resistance is anything other than 50Ω or if the load impedance shows any reactance, the line will not be perfectly matched. A mismatched line will not deliver all of the available power to the load. Some power will be reflected back down the line towards the generator. The two most extreme cases of a mismatched line occur when a line is terminated by a short circuit or an open circuit. The short circuit is an infinitely low resistance so it is not possible to generate any voltage across it. The open circuit is an infinitely high resistance so that it is not possible to pass any current through it. This means that no power is absorbed by the load and all of the available power is reflected back down the line. The interaction of forward and reflected power causes voltage and current peaks and nulls at half-wave intervals along the line. Many references to standing waves use the analogy of a travelling wave on a rope or a wave in a tank filled with water to illustrate the effect of standing waves on a transmission line.

VSWR (voltage standing wave ratio) is the most commonly used measure of the degree of mismatch on a transmission line. VSWR is the ratio of the peak voltage to minimum voltage along the length of the line. We will consider the extreme case of a lossless line terminated by a short circuit. All of the power available at the end of the line is reflected back down the line. Since our theoretical line is completely lossless, the peak voltage of the reflected wave is exactly the same as the peak voltage of the forward (incident) wave of the next cycle coming up the line from the generator. Constructive interference of the two waves will lead to voltage peaks which are twice the voltage of the incident wave and destructive interference of the two waves will produce voltage nulls of zero volts. This is a voltage ratio of 2/0 or infinity, the greatest possible degree of mismatch. A lossless line which is properly terminated will have no reflected power and the average voltage will be equal at all points along the line. This is a voltage ratio of 1:1 and therefore represents an SWR of 1:1. SWR can also be represented as ISWR (current standing wave ratio) which is the ratio of maximum to minimum current along the line.

MEASURING SWR. It is not easy to directly measure VSWR at short wave frequencies. As the voltage and current nodes and anti-nodes occur at different points along the line, it will be necessary to measure the voltage at one point on the line and at another point which is ¼wavelength away from this point. A ¼ wavelength on the lower frequency short wave bands is a distance of several tens of metres.

It is not always necessary to make a direct measurement of VSWR. In most cases, all we really need to know is that the impedance seen at one end of the line is equal to 50Ω or some other preferred value. This impedance is easily measured using a simple bridge circuit. A simple resistive bridge is shown in

Figure 5.17. This circuit is named after Charles Wheatstone who invented the rheostat and popularised the use of resistive bridges as measuring instruments. The device was actually invented by Samuel Hunter Christie in 1833. Wheatstone never claimed to have invented the bridge circuit and always gave credit to Christie for his invention.

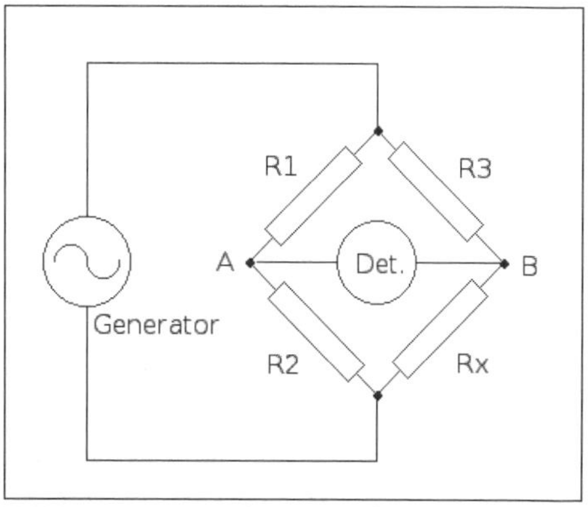

FIGURE 5.17: The classic Wheatstone Bridge.

To use a Wheatstone bridge as a simple 50 ohm

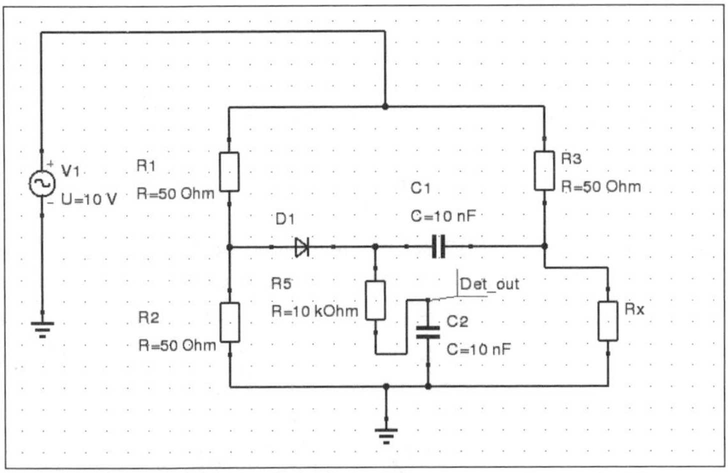

FIGURE 5.18: QUCS simulation of a practical 50W bridge.

SWR meter, R1 and R2 would be equal in value and a low inductance 50 ohm resistor would be used for R3. The voltage at point A is always exactly half of the generator voltage. The voltage at point B will only be the same as the voltage at point A if the unknown impedance Rx is a pure resistance which has the same value as R3. A sensitive detector is used to detect the voltage null when the bridge is balanced. **Figure 5.18** shows a QUCS simulation of a practical 50 ohm bridge. A radio transmitter is used as the generator and the aerial is connected to the Rx port of the bridge. A simple diode detector is used to measure the voltage across the bridge. Diodes with a low forward voltage drop will give more accurate measurements at low power levels. Point-contact diodes like the 1N34 or Schottky signal diodes are ideal for this purpose. **Photo 5.14** shows an RF bridge based on the circuit in **Figure 5.19**. The detector diode is a BAT43 Schottky diode (Maplin order code VR19V). R1, R2 and R3 are each made from a parallel pair of 1% tolerance 100 ohm metal-film resistors (Maplin M100R). This simple bridge is very easy to make and gives very accurate results at reasonably low values of SWR. It works well from LF to VHF and beyond. If low inductance SMT resistors are used, a simple resistive bridge will give accurate readings well into the microwave region **[63]**. Many sophisticated measuring instruments like SWR analysers, RF impedance bridges, noise bridges and return-loss bridges are based on simple resistive bridge circuits like the one shown in Figure 5.18.

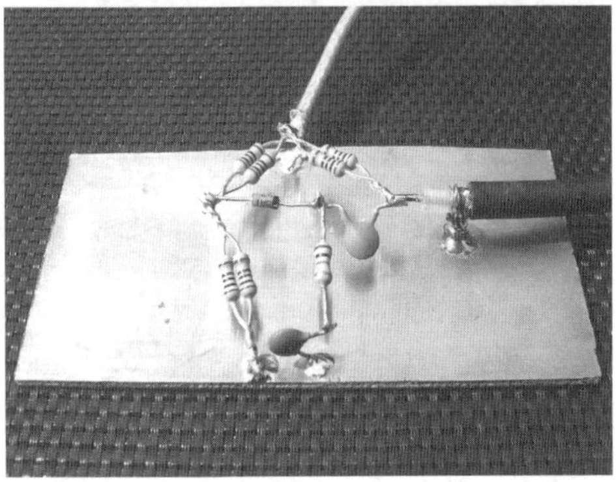

PHOTO 5.14: Practical implementation of a resistive bridge.

There are a couple of disadvantages to using a resistive bridge as an in-line SWR meter. When the bridge is balanced (SWR=1:1), the power from the transmitter is divided equally between the four arms of the bridge. This means that only 25% of the transmitter power will be delivered to the aerial feedline. This problem is easily overcome by switching the bridge out of line for normal operation. The bridge is only switched in when you want to check the SWR or re-tune the ATU.

The 6dB drop in transmitted power while tuning will give a welcome reduction in the amount of QRM on the amateur bands. The maximum power rating of the bridge will be limited by the power dissipation rating of the resistors. In the best case scenario where Rx=50Ω, each resistor will dissipate 25% of the transmitter power. If there is a mismatch at the Rx port, the dissipation in some of the resistors could be significantly more. An open circuit at Rx would present a 100Ω load to the transmitter and apart from the tiny amount of current drawn by the detector, R1 and R2 would have to dissipate all of the available power. A short circuit at Rx would create an even worse situation where R3 would have to dissipate 66% of the available power. The resistors used in my simple bridge have a total dissipation rating of 0.6W. Since there is a pair of resistors in each arm of the bridge, the maximum dissipation for a balanced bridge is 4.8W and considerably less when the SWR is high. In any case, the bridge will withstand brief transmissions of several watts without suffering any damage. The maximum power rating of the instrument can be increased by putting a 50Ω attenuator between the transmitter and the bridge. This will allow us to increase the maximum power rating to any level that is required, provided that the resistors in the attenuator are capable of dissipating the extra power. Resistive bridges are widely used by QRP operators. The limited power handling of a resistive bridge is not so much of a problem when the maximum transmitter power is only a few watts.

OTHER TYPES OF SWR METERS. Another instrument that can be used to determine SWR is known as the reflectometer. This instrument uses a pair of coupled transmission lines. The main transmission line runs through the instrument from the input connector to the output connector. The second line is a sampling line which is loosely coupled to the main line. This type of device is known as a coupler and is often named according to the degree of coupling between the main line and the sampling line, eg a 20dB coupler, 30dB coupler and so on. When one end of the sampling line is terminated in a resistance that is equal to its characteristic impedance, the device becomes directional, in that it only produces an output for power that is flowing in one direction. Such a device is known as a directional coupler. **Figure 5.19A** shows a directional coupler which is made from a section of co-axial line. One end of the sampling line is terminated by a resistor which is equal to its characteristic impedance. RF power at the other end of the sampling line is measured by a detector in the same manner as the resistive bridge above. It worth noting that there is no requirement for the coupling line to have the same characteristic impedance as the main line. **Figure 5.19B** shows a practical directional coupler for use in a VHF SWR meter. Two sampling lines are used so that it is possible to sample both forward and reflected power without having to reconfigure the circuit. This arrangement also allows for real-time calculation of SWR. This type of coupler is frequency sensitive so it can only be

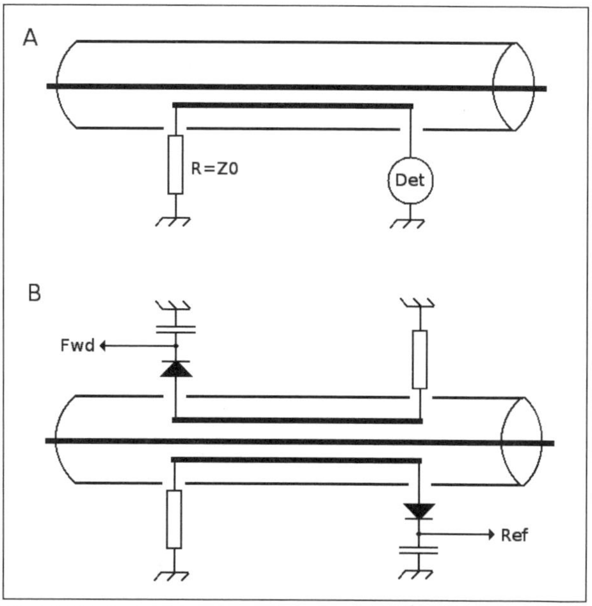

FIGURE 5.19 (A) Directional coupler made from a section of co-axial line (B) Practical directional coupler for use in a VHF SWR meter.

used for absolute power measurements if the meter is separately calibrated for each frequency band. Back in the good old days when many amateur stations used 75Ω co-ax and Belling-Lee TV plugs and sockets, directional couplers were often made by threading a wire through the holes in the dielectric of semi-air-spaced TV co-ax. Most types of 50Ω co-ax use a solid dielectric and are not easily used for this purpose. Fortunately, there are a few exceptions to this rule: some types of low-loss 50Ω co-ax like Westflex 103 have a semi-air-spaced dielectric. **Photo 5.15** shows a homebrew directional coupler made from a short length of Westflex 103. Two sampling lines are threaded through the coupler, one to measure forward power, the other to measure reflected power. Because the sampling lines are made from much thinner wire than the centre conductor of the main line, the characteristic impedance of the coupling lines is more than 50 ohms. Doing a few simple tests with the 2m transverter and a 50Ω dummy load shows that the impedance of the coupling lines is 86Ω. I used a pair of 200Ω skeleton pots to terminate the coupling lines for the tests. The 200Ω pots were adjusted for the deepest null in the reflected position and then the resistance of the pots was measured with a digital multi-meter. The main input and output connections were reversed several times during the tests so that both resistors could be adjusted for a deep null. After the tests were completed, the resistance of both pots was identical to within less than one ohm, indicating that the coupler is very well balanced.

PHOTO 5.15: Homebrew directional coupler made from Westflex 103.

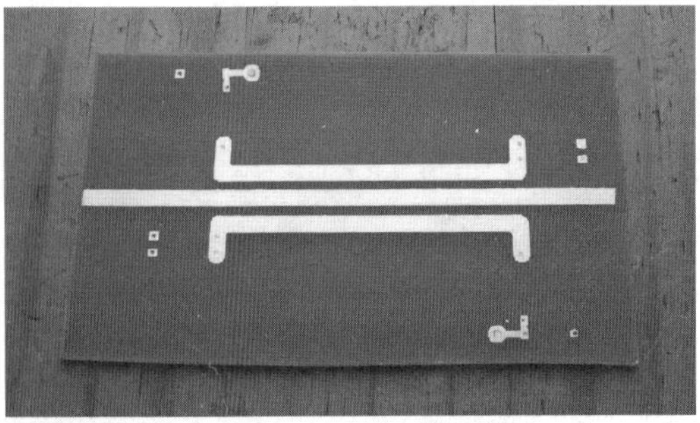

PHOTO 5.16: Home made directional coupler fabricated from 1.6mm double-sided PCB.

PHOTO 5.17: the completed directional watt-meter.

It is possible to make a directional coupler by using the copper tracks on double-sided PCB as coupled transmission lines. This is a cheap and easy way of mass-producing directional couplers. **Photo 5.16** shows a home made directional coupler which was made from 1.6mm double-sided PCB.

CONSTRUCTIONAL PROJECT. Our construction project for this month is a simple HF SWR meter / directional watt-meter. The directional coupler used in this meter uses a capacitive voltage divider to sample the voltage on the transmission line and a transformer to sample current on the line. The primary winding of the current transformer T1 is formed by a straight length of miniature 50 ohm co-ax cable which passes through the centre of a ferrite toroid. The secondary of the current transformer is 10 turns of bifilar wound enamelled copper wire wound on the toroid. Note that the shield of the miniature co-ax is only connected to ground at one side. The other side of the shield is left unconnected to form an electrostatic shield between the transformer primary and secondary windings. The toroid core is type 43 ferrite or similar. This circuit is reproducible and very easy to build. I have made several directional watt-meters using this circuit and all of them have performed very well. D1 and D2 are BAT43 Schottky diodes as used in the resistive bridge described above. A 300μA meter is used to show the output power and SWR. This is a broadband circuit with a very flat response from 1.8MHz to 30MHz so that forward and reflected power can be calibrated directly in watts. If a pair of meters is available, the forward/reflected switch can be omitted and both meters can be used to display forward and reflected power simultaneously. I used a 4-15pF ceramic trimmer for the bridge balancing capacitor. This capacitor will see quite high RF voltages across it at high power levels. Miniature trimmer capacitors might not be up to the job. Use a larger ceramic type or preferably, an air-spaced trimmer. **Figure 5.20** shows the schematic of the directional watt-meter. **Photo 5.17** shows the completed project.

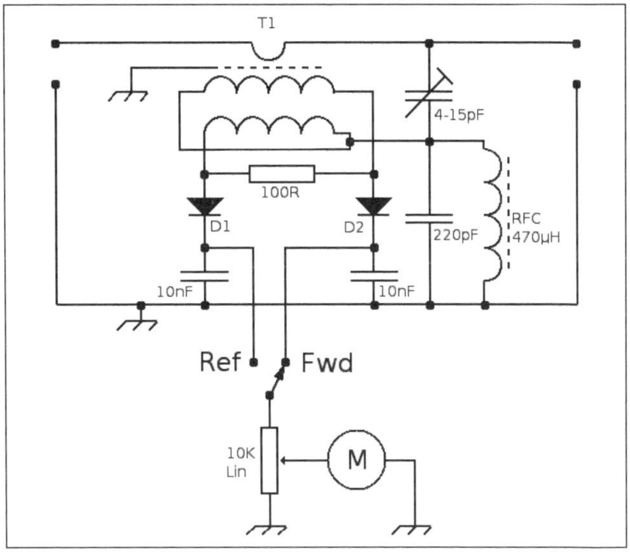

FIGURE 5.20: Schematic of a directional watt-meter.

Directional Power Meters

DIRECTIONAL POWER METERS. The HF directional power meter described previously is based on a Bruene **[64]** type of directional coupler. The Bruene circuit uses a transformer to sample RF current on a transmission line and a voltage divider to sample the voltage across the line. **Figure 5.21 (overleaf)** shows a simplified schematic of a Bruene coupler. The primary of the current transformer (T) usually consists of a single wire through the centre of a ferrite or powdered iron toroid core. RF current in the secondary winding develops a voltage across R3. This voltage at point B is proportional to the current on the line. The voltage divider resistors R1 and R2 produce a voltage at point A which is proportional to the voltage across the line. When a transmission line is terminated by a resistance which is exactly equal to the line characteristic impedance, the SWR on the line is 1:1 and voltage on the line is exactly in phase with line current. The values of R1 and R2 are chosen so that for a given value of line impedance (usually 50Ω), the voltage at A is equal to and in phase with the voltage at point B. This means that there will be no

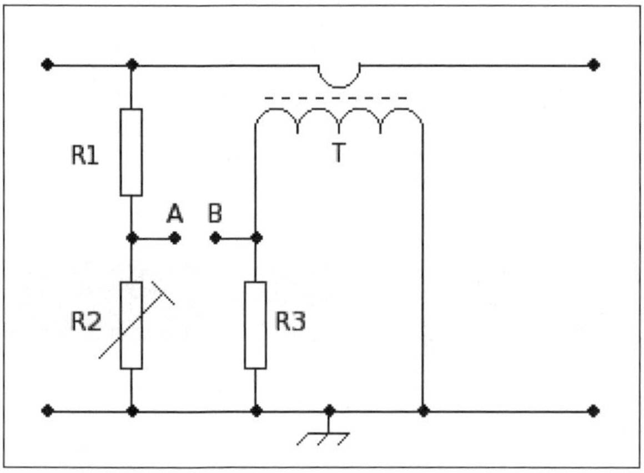

FIGURE 5.21: The basic Bruene directional coupler.

voltage across AB. This circuit is essentially the same as the Wheatstone bridge from last month's Homebrew. If the line is not properly terminated, reflected waves on the line will lead to out-of-phase voltage and current on the line which will unbalance the bridge and cause a voltage difference across AB. A sensitive detector at AB can be used to measure reflected power on the line. If the input and output connections are reversed, the bridge will indicate forward instead of reflected power. Swapping the input/output connections is a very inconvenient way of switching between forward and reflected power - it is much easier to reverse the polarity of the current sample by using a double-pole switch. If the secondary winding of the current transformer is centre tapped, a single pole switch can be used instead.

Figure 5.22 shows a directional coupler using a centre tapped current transformer and a capacitive voltage divider in place of the resistive divider that was used in Figure 5.21. The current transformer in our power meter project uses a centre tapped secondary and separate detector diodes (D1 and D2) to detect forward and reflected voltage. See Figure 5.20 for details.

CONSTRUCTION. The SWR/power meter was built in a small aluminium box (Maplin AB24 or similar). The input/output connectors are SO239 types. Square sockets with four holes for the mounting screws (Maplin BW85G) are preferable to the round type which are secured by a single large nut. **Photo 5.18** shows the method of construction. The original version of the meter used one pole of a DPDT switch for fwd/ref selection and a 10k pot to calibrate the meter for the power level in use. You can use a pair of meters to give a simultaneous indication of forward and reflected power as shown in **Figure 5.23**. R1 and R2 can be variable for SWR measurement or pre-set for forward/reflected power measurement.

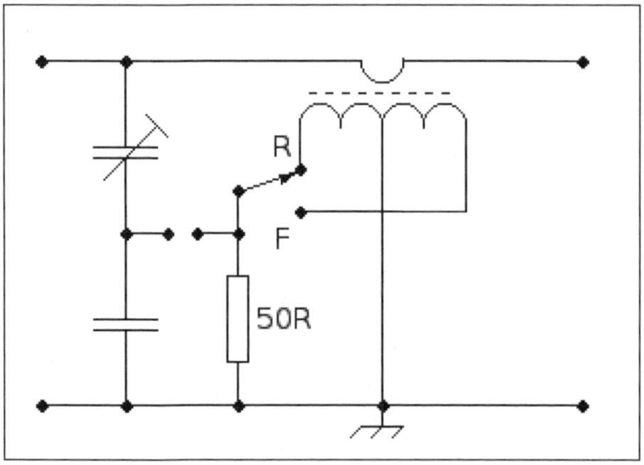

FIGURE 5.22: The Bruene circuit modified to read forward and reflected power.

TESTING AND CALIBRATION. The instrument should be placed in line between a transmitter and a good quality dummy load. The transmitter should be connected to the TX connector and a good quality 50 ohm dummy load should be connected to the aerial connector. Use 50Ω co-axial cable and connectors to make the connections.

Set the meter calibration pot to about 25% above the minimum position. Set the transmit frequency to the middle of the transmitter frequency range. (I did the initial calibration at 14MHz). Do the initial tests at a reasonably low output level of about 5-10W. **Figure 5.24** shows the very simple dummy load and peak voltmeter that was used to measure the transmitter output power. The three resistors are 150Ω, 2W metal film types. The diode is a Schottky signal diode, BAT43 or similar. The power in watts is the RMS voltage squared, divided by the load resistance. $(V\_pk*0.7071)^2/50$. Add 0.25V to the measured voltage to compensate for the forward voltage drop of the Schottky diode, ie if the DC voltmeter shows that the peak voltage is 31.37 and we add 0.25V for the diode, the actual voltage is 31.62V. Using the above calculation,

$P = (31.62 * 0.7071)^2 / 50$
$P = 9.998W$

or as near to 10W as to make no difference

The Fwd/Ref switch should be set to the Fwd position. At this stage, I had no idea which side of the switch indicated forward power, so I simply selected the side that gave the higher power indication. Set the meter calibration pot so that the meter gives a full scale indication at the test power level. The Fwd/Ref switch should then be moved to the reflected position and the trimmer capacitor in the voltage divider network adjusted for the lowest possible reading of the meter. Use a small insulated screwdriver for making this adjustment. A screwdriver with poor insulation could lead to RF burns at higher power levels. Take care with this adjustment, repeat the adjustment several times and try to find the deepest possible null. The stray capacitance of the screwdriver will interact with the adjustment so you will have to make several adjustments to find the best result.

PHOTO 5.18: Method of constructing the SWR/power meter.

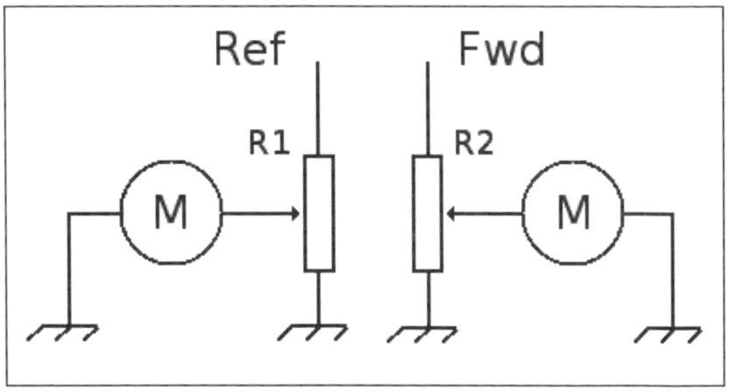

FIGURE 5.23: How the meters are arranged for simultaneous forward and reverse indication.

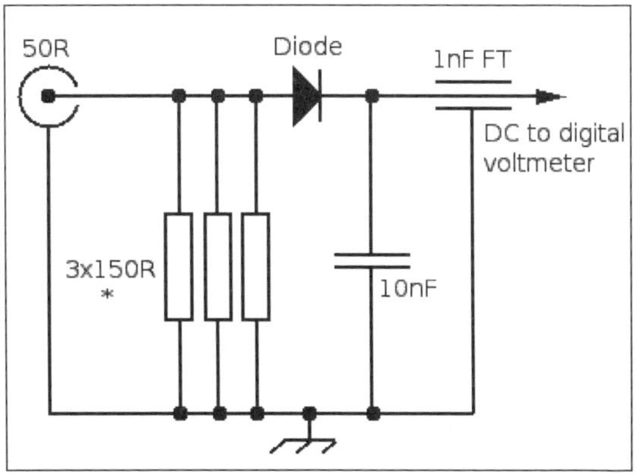

FIGURE 5.24: The simple test load used to calibrate the prototype.

Once the bridge has been carefully balanced for a 50Ω line, no further adjustments to the directional coupler will be necessary. Before you calibrate the power scale of the meter, you should decide what the full-scale power level will be. My transmitter can produce more than 200W on the LF bands and about 170W on 10m. I decided to calibrate the meter so that the full scale reading is 400W.

PHOTO 5.19: The SWR/power meter with hand-calibrated scale.

Because the meter follows a square law, the meter shows 100W at half scale, 25W at quarter scale and so on.

There are two ways of approaching the calibration process. You can draw a square law scale on the meter and place your trust in the linearity of the meter – or you can take a more methodical approach by testing the meter at various power levels and then calibrating the scale accordingly. I chose the latter option. A good quality micro-ammeter will show a nice linear relationship between current and meter deflection. Cheap low quality meters like the one I used can be quite non-linear.

I used a high power dummy load and peak reading voltmeter to make accurate measurements of the transmitter output. The meter reading on the original meter scale was noted at various power levels. I checked the reading at 25W, 50W, 100W and 200W. The original scale was then removed from the meter and replaced by a new scale. If you have a steady hand and a keen eye, you could draw the new scale by hand. Since I have neither, I used a computer drawing program (Xfig) to create the new scale. **Photo 5.19** shows the meter.

I used a handful of 2W metal film resistors to check the calibration of the SWR scale. A 150Ω resistor should give an SWR of 3:1, which should show about half scale on the meter. A 100Ω resistor will give 2:1, a parallel pair of 150Ω (75Ω) gives 1.5:1 and a parallel pair of 100Ω gives a perfect 1:1. Keep your test transmissions short, because putting 5-10W into a 2W resistor will produce a lot of smoke if you transmit for more than a few seconds!

Once the calibration was checked on 20m, I tested the power and SWR accuracy on all nine HF bands. I was very pleased and more than a little surprised to find that the meter read within a few percent of the expected reading on every band from 160m to 10m. I didn't have a 6m transmitter to hand so no tests were performed on that band. The formula for converting forward/reflected power to reflection-coefficient and VSWR is as follows:

ρ (Rho) = Reflection Coefficient

$\rho = \sqrt{Pref / Pfwd}$

$VSWR = 1 + \rho / 1 - \rho$

$\rho = VSWR - 1 / VSWR + 1$

A QRO DIRECTIONAL POWER METER. The other type of broadband directional coupler in common use is often referred to as the 'Tandem Match' **[65]**. This circuit uses a pair of broadband transformers, one to sample current on the line and the other to sample voltage across the line. **Figure 5.25** shows a schematic of the Tandem Match type of coupler. The two transformers T1 and T2 are identical. T1 samples current on the main line through the instrument and T2 samples voltage across the main line. A secondary line which runs through T2 is terminated at both ends by resistors R1 and R2. This type of directional coupler needs no adjustment. The value of R1 and R2 is equal to the characteristic impedance of the main line. To make a 50Ω bridge, use 50Ω resistors for R1 and R2. To make a 75Ω bridge, simply change the value of these resistors to 75Ω.

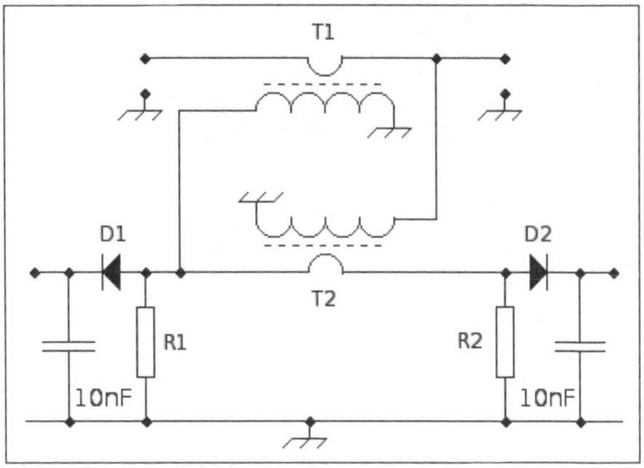

FIGURE 5.25: Schematic of the Tandem Match coupler.

Figure 5.26 shows a QUCS simulation of a Tandem Match coupler. V1 is an AC voltage source with a peak (not peak-to-peak) voltage of 100V and a frequency of 3.7MHz. TR1 and TR2 are broadband transformers which are used as 30dB couplers. The primary is 1 turn through a ferrite toroid, the secondary is 32 turns.

The coupling factor is 20log(N), where N is the transformer turns ratio, so 20log(32)=30.1dB. **Figure 5.27** shows a QUCS voltage/time plot of the output from the forward and reflected ports of the coupler. The red trace shows forward voltage of 100/32=3.125V and the blue trace shows reflected voltage very close to zero volts.

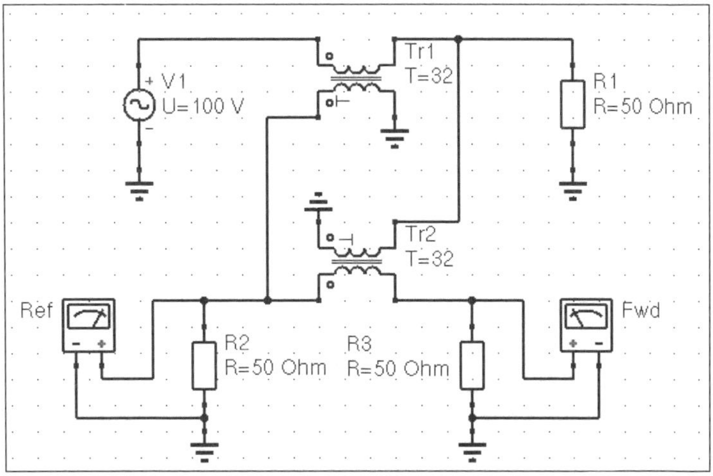

FIGURE 5.26: QUCS simulation circuit for a Tandem Match coupler.

CONSTRUCTION. The first version of the tandem match used a pair of 26dB couplers (20 turn secondary) wound on FT37-43 cores. This arrangement worked well from 80m to 10m. I found that the voltage sampling transformer got very hot when used at 200W+ on the low end of the 160m band. The final version uses a pair of 30dB couplers wound on large ferrite cores. The cores are Maplin N88AB RFI suppression toroids. The primary winding is a length of Westflex 103 or similar co-ax. As with the Bruene coupler described earlier, the outer conductor of the co-ax is only grounded at one end so that it acts as an electrostatic shield between the primary and secondary of T1 and T2. It is very important that the construction of the two transformers is absolutely identical. The secondary winding of the

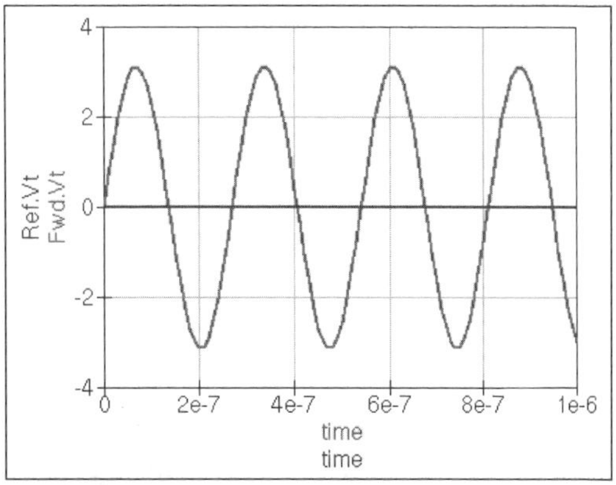

FIGURE 5.27: QUCS simulation result for the Tandem Match coupler.

PHOTO 5.20: The finished Tandem Match circuit.

transformers is 32 turns of insulated wire. I used two lengths of wire from a length of standard 4 core telephone cable, with a small blob of epoxy resin to secure the toroid core in position. **Photo 5.20** shows the finished Tandem Match circuit. This directional coupler shows a very flat frequency response from 1.8 to 30MHz and runs stone cold at 250W on 160m. It should be capable of coping with power levels well above the UK legal limit. The 50 ohm terminating resistors for the coupled ports (R1, R2) are made from a parallel pair of 100 ohm, 1% tolerance metal film resistors (Maplin M100R or similar).

Calibration of the Tandem Match is a very simple affair. There is no way of adjusting the balance of the bridge. Provided that T1 and T2 are identical in construction and R1/R2 provide a good low inductance 50Ω load for the coupled ports, you should find that the bridge is well balanced. Note that this bridge is designed for relatively high power operation. Non-linearity in the diode detector tends to lead to inaccurate readings at very low power levels. The forward voltage drop of the BAT43 diode is about 0.25V at low current levels. This means that the detector will produce little or no output at power levels below 640mW (calculation: 0.25*32=8V; $(8*0.7071)^2/50=0.640$W). The meter is very accurate at power levels above 2.5W (0.5V at the detector).

I used a 100W transmitter and dummy load to calibrate the power scale of the meter. The configuration of the test equipment was similar to the system used for calibrating the Bruene coupler except for the higher transmitter power and the much higher power rating of the dummy load. The dummy load consisted of a pair of 100Ω, 30W resistors which were bolted to a substantial heatsink. This load proved perfectly capable of withstanding more than 100W if the transmission periods were limited to less than 10 seconds for each test. The single Schottky diode in the peak voltage probe was replaced by three 1N4148 diodes connected in series. I allowed for a total voltage drop of 1.8V for three diodes when calculating the peak voltage across the load. An indicated voltage of 98.2V on my digital voltmeter shows that the peak voltage across the 50Ω load is 100V. $(100*0.7071)^2/50=100$W. I also used the 400W square law scale from the Bruene meter for this project. 100W forward power and 3:1 SWR are at the 50% point on the scale. I tested the meter accuracy at several points on the scale on all nine HF bands. Power accuracy was within a few percent on all bands. The meter showed a flat 1:1 SWR from 160m to 17m. There was a very slight reflected power indication (SWR = 1.1:1) on 10m. I don't know whether this was due to stray inductance in my dummy load or a slight imbalance in the coupler. The power and SWR calibration tests were repeated with the input and output connections of the bridge reversed to ensure that the bridge was well balanced. This resulted in a reflected power reading of exactly 100W and a forward power of 0W indicating that the bridge is well balanced and symmetrical.

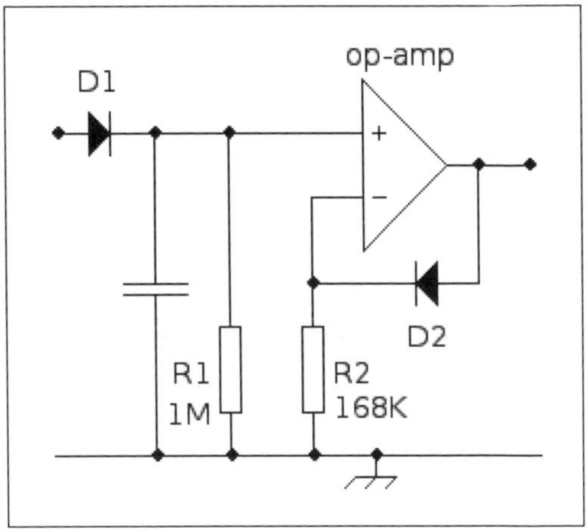

FIGURE 5.28: Using an op-amp and compensating diode can give accurate results down to the milliwatt range.

COMPENSATING FOR DETECTOR ERRORS. Diode detectors tend to become very non-linear when measuring low voltages. Silicon junction diodes have a forward voltage drop of about 0.6V. Schottky diodes and germanium point-contact diodes have a forward voltage drop of about 0.2-0.3V at low current levels. This small voltage drop is usually tolerable when the detected voltage is several volts or more. The effect of non-linearity in diode detectors is magnified when they are used to measure the output of voltage dividers or couplers as used in the power meters above. There are a number of ways of compensating for errors in diode detectors. You can use a broadband linear amplifier between the output of the coupler and the detector. This will increase the voltage to a level where the small voltage drop of the diode becomes insignificant. The amplifier must have a very flat frequency response across the entire frequency range of the meter and sufficient dynamic range so that it remains linear for the full power range of the meter. Some power meters use a log amplifier to measure the output from the directional coupler. Log amplifiers like the Analog Devices AD8307 are used for power measurement in instruments like spectrum analysers, radar systems, RSSI (received signal strength indicators) or any application that calls for logarithmic output and a high dynamic range. The AD8307 had a dynamic range of 92dB and accuracy of +/-1dB from LF to VHF **[66]**. The AD8307 is used in several published power meter designs, including the excellent N2PK power meter **[67]**.

An op-amp can be used to compensate for the forward voltage drop of the detector diode. The most common arrangement uses a pair of matched diodes. One diode is used as the detector and the other diode is used as a reference. Provided that the voltage drop of both diodes is identical, the op-amp can almost completely compensate for the voltage drop of the detector diode. **Figure 5.28** shows the schematic of a diode detector with op-amp compensation. This type of detector allows for accurate power measurements right down to the mW region.

D1 is the detector diode, and D2 is an identical diode which is used as a reference. The current through D1 depends on the value of the 1M load resistor and to a lesser extent on the input current of the op-amp. Because D1 has a capacitive load and it is driven from an AC source, the peak current through D1 is much higher than the average current. The value of R2 is much less than the value of R1 to compensate for this imbalance [65]. The op-amp should have a very high input impedance and low input offset. MOSFET input types like the CA3140 are well suited to this circuit. The inputs and output of the CA3140 can operate all the way down to the negative supply rail which allows this circuit to be used with a single-ended power supply (no negative supply needed). This circuit will feature in a precision dummy load and power meter in the near future.

Grid Dip Oscillator

A CENTURY ON AND STILL USEFUL. A hundred years after the invention of the triode valve, the grid dip oscillator (GDO) is still one of the most useful instruments in the amateur radio shack. The valve has been replaced by the FET and other solid-state devices, but the basic GDO circuit remains much the same as it was in the 1920s. **Figure 5.29A** shows a schematic of a GDO based on a triode valve. **Figure 5.29B** shows a solid state version of the same circuit based on a JFET. The theory behind this circuit is quite simple. The triode is configured as a Colpitts oscillator. The frequency of oscillation is determined by the values of C1 and L1. Positive feedback is fed from the anode circuit of the valve back to the grid circuit via a centre tapped dual gang capacitor (C1). The inductor L1 is mounted externally so that it can be placed in close proximity to the circuit under test. When L1 is loosely coupled to an external resonant circuit, a dip in grid current (or gate current in the case of the FET) can be measured when the grid-dip oscillator is tuned to the resonant frequency of the external circuit.

The name grid dip oscillator is a bit of a misnomer for modern dip oscillators. Solid state dip oscillators are based on transistors, FETs or special negative resistance devices like the tunnel diode. GDO is not an appropriate name for a dip oscillator based on a bipolar transistor because the control input of a bipolar transistor is called the base. As the control input of a FET (field effect transistor) is called the gate, dip oscillators based on FETs are often referred to as gate dip oscillators.

The classic GDO design is a hand held instrument with a variable capacitor for tuning the oscillator frequency. Plug in coils are used so that the instrument can be used over a wide range of different frequencies. The oscillator frequency is read from a calibrated scale on the front of the instrument. A separate scale is required for each plug in coil. **Figure 5.30** shows a typical arrangement of a dip oscillator. Some of the older designs need a mains power supply to provide heater current and HT for the valve.

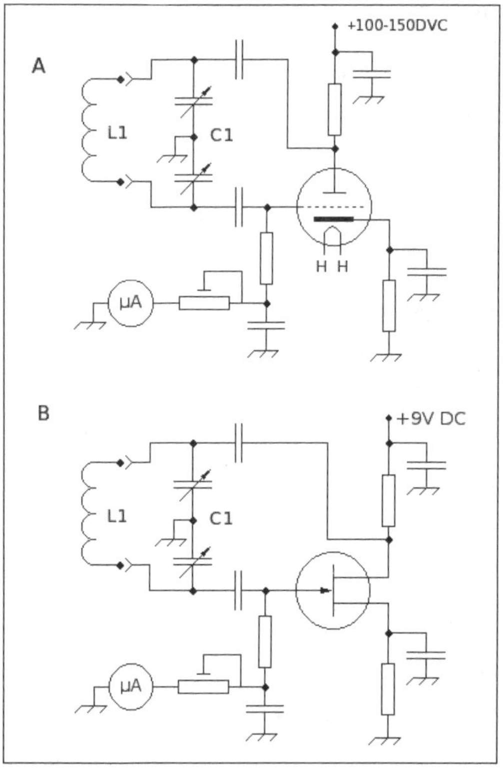

FIGURE 5.29: 'Grid dip' oscillator circuits (A) with a triode (B) with a JFET

FIGURE 5.30: How a typical GDO is used.

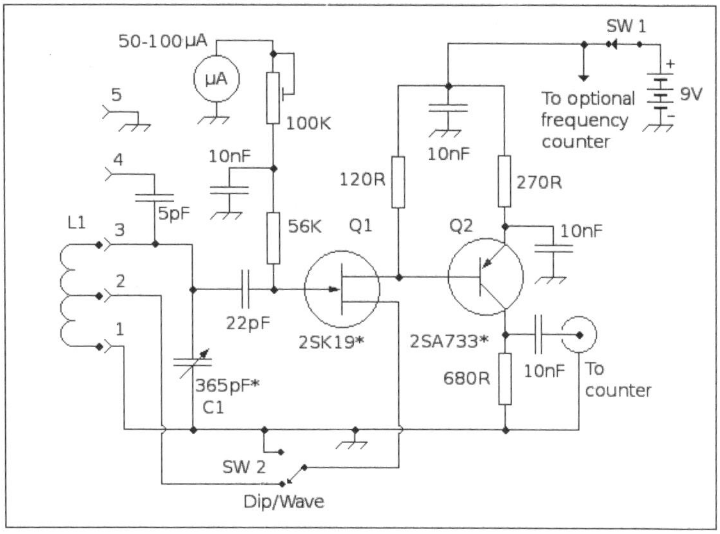

FIGURE 5.31: Practical GDO circuit diagram .

Most solid state dip oscillators are powered by a battery. The use of a battery as the power supply is a major advantage for a hand held instrument. Battery powered dip oscillators are safe and easy to use outdoors. Since a transistor or FET oscillator will only draw about 10mA from the power supply, you can expect to get many hours of operation from a small 9V battery.

THE CIRCUIT. This month's project is an HF GDO which can also be used as an absorption wave-meter. The circuit of the dip oscillator is shown in **Figure 5.31**. Q1 is a junction FET which is configured as a Hartley oscillator. I used the Hartley configuration instead of the more commonly used Colpitts configuration because I didn't have a suitable dual-gang capacitor for C1. A single 365pF variable capacitor was used for C1. The FET is connected in the common drain configuration instead of the common source configuration used in the earlier example. This circuit allows one side of the tuning capacitor to be connected directly to ground. Positive feedback is fed from source to gate via a tapped inductor L1. Gate current is measured by a microammeter connected in series with the gate resistor. A 50μA or 100μA panel meter is ideal for this job. Q2 is a PNP transistor which acts as a buffer amplifier between the oscillator and the frequency counter output. This output can be used to drive the optional internal frequency counter or a separate external frequency counter. SW2 is used to kill the positive feedback so that Q1 acts as an amplifier instead of an oscillator. This allows the instrument to be used as a tunable absorption wave-meter and a very sensitive frequency counter. The terminal marked '4' on the left side of the schematic is for connecting a capacitive probe to the dip meter as an alternative to inductive coupling to L1.

The oscillator frequency depends on the values of L1 and C1 and to some extent on the input capacitance of the FET and other stray capacitances. The well known formula for LC circuit resonance is $f=1/((2*\pi)*\sqrt{(L*C)})$, where f is frequency in Hz, L is inductance in Henries and C is capacitance in Farads. The formula can be rewritten as $f=1000/((2*\pi)*\sqrt{(L*C)})$, which gives the resonant frequency in MHz if L is in μH and C is in pF.

CONSTRUCTION. This is an ideal first project for the novice constructor of test equipment. The oscillator and buffer amplifier at the core of this project is a very simple circuit and should be easy to reproduce. The circuit was built on the copper side of some PCB laminate using dead-bug style point-to-point wiring. I used a 2SK19 JFET for Q1 and a 2SC733 PNP transistor for Q2. Most general purpose N-channel FETs should work well in this circuit. I also had good results with a BF245C and a J113 FET. The only component which might be difficult to find is the variable capacitor C1. I used a small receiving type of air-spaced capacitor with a nominal maximum capacitance of 365pF. The measured maximum capacitance is closer to 400pF and the measured minimum capacitance is about 25pF. A miniature plastic film dielectric 'Polyvaricon' type of variable capacitor can also be used in this circuit. L1 is a home made plug in coil. More details about this coil later.

The last time I built a GDO was more than twenty years ago. The oscillator section of that GDO was quite similar to the one used in this new project. The initial construction of the previous GDO was very easy, but the design of the plug in coils and the calibration of the frequency scale turned out to be a bit of a nightmare. The need to draw several different frequency scales by hand (one for each coil) was a particularly tedious job. This time, I decided to take a different approach. The new GDO/wave-meter includes a built-in digital frequency counter. This means that there is little or no calibration required. The frequency counter also makes it very easy to make new plug-in coils. All you need to do is calculate the approximate value of inductance required for the frequency range that you want to cover, wind a new inductor with slightly more turns than are required and if necessary remove a few turns to get the oscillator on exactly the right frequency. As the built in counter gives a real time indication of the oscillator frequency, it tales no more than a few seconds to check the oscillator frequency at both ends of the capacitor's tuning range.

The simple frequency counter was described earlier in this book. No modifications to the counter hardware were required for use in the GDO project. However, I did make some modifications to the counter software. The original version of the counter firmware used a count gate time of 1 second. This is far too slow to track the rapidly changing frequency of a variable frequency oscillator. The modified PIC program for the counter uses a gate time of 100ms. This reduces the count resolution to 10Hz instead of the original 1Hz. This is of no practical significance since the free running oscillator has a typical drift rate of several Hz per second. The modified version of the counter firmware can be found at: http://tinyurl.com/q24xj. The counter unit takes its power from the 9V battery.

The project is built in a small aluminium box (Maplin AB24 or similar). Despite the fact that it makes the end result look rather ugly, I decided to build the entire project into the bottom half of the box. This means that the whole project has an upside-down appearance. The LCD module and the tuning capacitor are mounted in the bottom of the box. This arrangement has the advantage of having all of the components mounted in the same half of the box. The alternative arrangement would place the frequency display and gate current meter in the top half of the box and the oscillator and coil socket in the bottom half of the box. This would undoubtedly look better, but it would be very inconvenient to have interconnecting wires between the GDO circuit in the bottom half of the enclosure and the components mounted on the removable lid. **Photo 5.21** shows the finished project. I used a nibbling tool to cut out the rectangular holes for the LCD display and microammeter. Take care at this stage, it is very easy to nibble away the sheet metal, but it is not so easy to put it back when you make a mistake.

COILS. Each of the plug-in coils for L1 were wound on a short length of plastic tubing. I found that the centre part of a 5 pin 180 degree DIN plug is a tight fit in the end of the most common type of reinforced garden hose. I cut several short lengths of tubing, one for each coil. I used a small amount of superglue to secure the DIN plugs in position at one end of each tube. See **Photo 5.22** for details. So far, I have made three coils which give continuous coverage from 1.6MHz to 36MHz. Coil 1 is 45 turns tapped at 15 turns, coil 2 is 21 turns tapped at 7 turns and coil 3 is 6 turns tapped at 3 turns. 0.5mm enamelled copper wire was used for all three coils. The diameter of each coil is approximately 15mm. The standard textbook formula predicts that an air cored coil of 45 turns using the dimensions for coil 1 above would have an inductance of about 15.3μH. I have never seen any published specifications for the initial permeability or AL values of green garden hose. To measure the value of coil 1, I connected it in parallel with a 47pF fixed capacitor and used coil 2 in the GDO to find its resonant frequency. A sharp dip was observed at 5.03MHz which indicates an inductance value of 21.3μH. The formula for finding the value of an unknown inductance is $L=25330/f^2*C$, where L is the inductance in μH, f is the resonant frequency in MHz and C is the value of the capacitor. $25330/5.03^2*47 = 21.3$μH. The inductance of coil 2 was measured at 6.6μH and coil 3 has a value of 0.85μH.

PHOTO 5.21: The prototype GDO.

PHOTO 5.22: The three prototype coils.

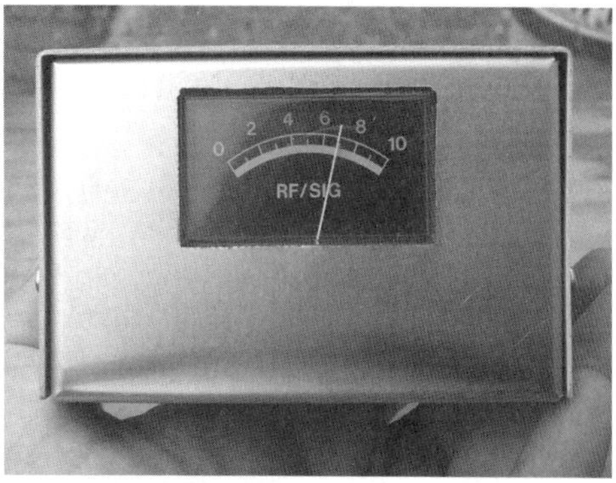

PHOTO 5.23: There is no need to calibrate the meter – use whatever scale it comes with.

The approximate tuning ranges with these coils are:

Coil 1: 1.6MHz to 7MHz
Coil 2: 3MHz to 14MHz
Coil 3: 8.3MHz to 36MHz

The rest of the construction is quite easy. I used four M2.5 screws and nuts to mount the LCD display module. I had to cut four stand-off spacers from 3mm brass tubing to give the correct spacing between the PCB of the display module and the aluminium box. The socket for the plug in coils is a standard 180 degree 5 pin DIN socket (Maplin HH34M or similar). The 100μA meter is held in place by a simple clamp made from a length of plastic bar and a couple of M3 screws. The battery is taped on to the back of the meter clamp. A SPDT switch is used to switch the battery power on and off and another SPDT switch is used to select dip-oscillator or wave-meter mode. **Photo 5.23** shows the microammeter. Since we are only interested in a relative dip in gate current rather than the absolute value of the gate current or voltage, I didn't make any attempt to calibrate the scale of the meter.

CALIBRATION AND TESTING. Plug coil 2 into the coil socket and set C1 to the middle of its range. Set the 100k pot so that the meter is showing 60-70% of full scale deflection. To calibrate the frequency counter, tune a receiver to a known good frequency reference and tune the GDO until it is zero beat with the reference. I usually use the atomic clock signal on 9.996MHz for this purpose. If necessary, adjust the crystal trimmer capacitor on the frequency counter PCB until the correct frequency is displayed. Alternatively, you can calibrate the frequency counter against another frequency counter or a frequency standard oscillator.

USING THE GDO. Building the dip oscillator is the easy part of the project. Learning how to use it properly and how to interpret the measurement results will probably take longer than the construction of the project. The dip meter can be used to determine the frequency of an external resonant circuit. **Photo 5.24** shows how the instrument can be used to find the resonant frequency of a parallel tuned circuit. The inductor is a junkbox component with unknown properties. The capacitor is a 100pF ceramic disc capacitor. The two wires running from the inductor to the capacitor form a closed loop which is an ideal location for coupling to the GDO. Careful tuning of the GDO reveals that there is a pronounced dip at 13.8MHz. This is the parallel resonant frequency of the external LC circuit. Beware of false or inconsistent readings. You should find that the dip disappears when the GDO is moved away from the circuit under test. Gradually increase the distance between the GDO coil and the circuit under test until the dip is barely perceptible. You will see that there is a significant gap between the GDO and the external circuit in Photo 5.24. Re-tune the GDO several times to ensure that you have found the deepest dip on the meter. The resonant frequency of the external circuit can now be read directly from the LCD frequency display.

Because of the self-shielding properties of toroidal inductors. It is sometimes difficult to achieve a sufficient level of inductive coupling between the test circuit and the GDO coil. This problem can be overcome by passing a single wire through the toroidal core and connecting the ends of this wire between the capacitive probe terminal and the ground terminal of the GDO.

Some test instruments like a multimeter or a digital counter will give a very clear indication of the test results. If the instrument is well calibrated, the reading will be accurate and the operator will have no difficulty in interpreting the results. Some other instruments like the oscilloscope or the GDO require a greater level of operator skill. This skill comes with practice. After you have spent some time experimenting with a few different LC circuits, you will develop the ability to find resonances in the test circuit quickly and accurately.

The GDO can also be used to find the resonant frequency of aerials. **Photo 5.25** shows how the instrument can be used to find the resonant frequency of a dipole aerial. The centre of the dipole is formed into a two turn loop to provide a convenient method of coupling the GDO to the aerial. This method can be used to find the resonant frequency of half-wave dipoles, quarter-wave monopoles and closed loop aerials like the cubical quad and delta loop.

The GDO can also be used to find the resonant frequency and velocity factor of transmission lines. Open or short circuit lengths of parallel wire transmission line or co-axial lines are often used as resonant stubs or matching transformers. The resonant frequency of an open circuit quarter-wave line or a short circuit half-wave line is easily measured with the GDO. **Photo 5.26** shows how the instrument can be used to measure the resonant frequency of a 75Ω co-axial line. The inner and outer conductors are formed into a closed loop at one end of the cable. The find a quarter wave resonance, the far end of the cable should be left open circuit. To find a half wave resonance, the far end of the cable should be shorted by connecting the inner conductor to the outer braid. The length of the test cable is 2140mm. A sharp dip was found at a frequency of 28.9MHz. This is the quarter-wave resonant frequency of the open circuit line. This type of resonator is known as a quarter-wave stub. Since we now know the resonant frequency of the co-ax line, we can easily calculate the velocity factor (VF) of the cable. First we must measure the physical length of the cable so that we can compare it to the electrical length of the cable.

PHOTO 5.24: Testing an unknown L-C circuit with the GDO.

The simple formula 75/f gives the length of a free space quarter wavelength in metres. 75/28.8 = 2.595 metres. To get the VF of the cable, simply divide the physical length of the cable by the free-space quarter wavelength. 2140/2595=0.82. The same test was carried out on various lengths of cable at several different frequencies and the VF always worked out at 0.82. The cable is RG6 satellite TV cable with a foam dielectric. I also tested a length of home made open wire 600 ohm line to see if it exhibits the often quoted VF of

PHOTO 5.25: Using the GDO to find the resonant frequency of a dipole aerial.

PHOTO 5.26: Using the GDO to measure the resonant frequency of a 75 ohm co-axial line.

PHOTO 5.27: Inside the prototype GDO.

0.95 to 0.97. My quarter-wave stub of open wire line has a velocity factor of 0.93. This small discrepancy may be due to the fact that my line is made from PVC insulated wire.

FINDING THE VALUE OF UNKNOWN INDUCTORS AND CAPACITORS. The GDO provides an easy way to find the value of unmarked capacitors and inductors. As we saw above, it is possible to accurately measure the resonant frequency of an LC resonant circuit. The GDO will tell us the resonant frequency. If we know the value of one of the reactive components, it is a simple matter to calculate the value of the other. In our earlier example, the resonant frequency was 13.8MHz and the capacitor had a known value of 100pF. $25330/13.8^2*100$ gives a value of 1.33μH for the inductor. If the value of the inductor is known, the similar formula $C=25330/f^2*L$ gives the value of the capacitor. $25330/13.8^2*1.33=100$pF.

If you have a very steady hand and a keen eye, it is possible to use the GDO to find the resonant frequency of quartz crystals. I tested a few crystals by connecting a two turn wire loop to the crystal pins and using the GDO to find the crystal resonance. If the approximate frequency of the crystal is already known, it is very easy to find the series resonance. Even with unmarked crystals, it is possible to find the resonance if you tune the GDO *very* slowly.

This instrument can also be used as an absorption wave-meter. Set the switch to the wave-meter position and place the instrument close to the circuit that you want to test. This is an excellent way of checking your transmitter for harmonics, IF image and other spurious outputs. Tune C1 to find the output frequency of your transmitter and then sweep the entire HF range for any sign of spurious signals. Take care when testing higher power transmitters. Too much power into the gate of the JFET could damage it.

References

[53] http://tinyurl.com/2jzbu2
[54] http://tinyurl.com/yrj5k2
[55] www.cebik.com/phase/hb.html
[56] www.iol.ie/~bravo
[57] *HF Antennas for All Locations*. L A Moxon, G6XN, 1982, RSGB, Chapter 12 Horizontal Beams
[58] www.cebik.com/moxon/moxgen.html
[59] www.ac6la.com/
[60] http://homepage.eircom.net/~ei9gq/beam.html
[61] www.ifwtech.co.uk/g3sek/diy-yagi/index.htm
[62] http://homepage.eircom.net/~ei9gq/atu.html
[63] http://pe2er.nl/wifiswr
[64] An Inside Picture of Directional Wattmeters. Warren B. Bruene W0TTK. QST April 1959.
[65] The Tandem Match – An Accurate Directional Wattmeter. John Grebenkemper KI6WX. ARRL Handbook 1993
[66] AD8307 data-sheet. Analog Devices 2006
[67] http://n2pk.com/#TP2

INDEX